REVIVE. ALIVE. THRIVE.

A PATH FOR TRADITIONAL BUSINESSES
TO STAY AHEAD WITH

DIGITAL
TRANSFORMATION

FAN WU

Published by Archive Zero, LLC

Paperback ISBN: 978-1-955722-03-2
Hardback ISBN: 978-1-955722-04-9
E-book ISBN: 978-1-955722-02-5

Cover design by Robson Garcia Jr.
Formatting by Polgarus Studio

Contents

REVIVE. ALIVE. THRIVE.

INTRODUCTION

LET'S START WITH TWO QUESTIONS...

IS YOUR BUSINESS DIGITAL?
IF YOUR BUSINESS IS NOT DIGITAL, WHAT KINDS OF BUSINESS CAN BE CALLED DIGITAL?

For the first question, you would probably quickly say no, that your business is ultimately about health care, transportation, automobiles, or fast-moving consumer goods such as food, beverage, beauty and fashion.

For the second question, you would probably say "high-tech companies", such as Google, Amazon, Facebook (rebranded as Meta), Apple, or even new players such as Uber, Airbnb, and Netflix.

You might believe digital is increasingly relevant for you, but that it is not absolutely critical. However, digitalization *is* critical for every business, in every industry, and everywhere in a new digital economy era. This technology is pervasive, yet most executives are the least knowledgeable about—and the most unprepared for—the transformation it can and will bring about.

DIGITAL IS CONSUMING THE PHYSICAL, INDUSTRIAL WORLD

The latest and greatest technologies on the horizon are highly discussed among technical people. You have no doubt been inundated with articles and presentations about cloud computing, social, mobile, and cognitive technologies. To that list, we can add mobile, robotics, block chain, artificial intelligence, 3D printing, drones, nanotechnology, virtual reality, augmented reality and mixed reality (virtual reality mixed with actual reality).

As usual, we love to talk about the Fortune 500. We love to debate what this annual published list with the highest revenues implies about economic trends. So when about fifty companies on the magazine's 1950 list, less than 10 percent, were still on the list in the second decade of the 21st century, what does this signify?

The reordering of the business leaderboard—represented by the Fortune 500—will increasingly be influenced by digital technologies. However, this shift is not about technologies themselves influencing the business world, but the mindset of a digital economy, as well as the benefits from new digital technologies, that accrue only when companies change their organizational architecture: their structure, process, talents, and decision-making orientation, including shareholder relationships, teamwork coordination, resource allocation, logic, Key Performance Indicator (KPI) measurement and incentive rewards.

Going digital embraces the new business infrastructure, one that is at the intersection of powerful computing, pervasive connectivity, and a potent cloud. In fact, three laws are key driving forces in the world of digital technology development. First is Moore's Law, which means computing power grows exponentially and at lower and lower cost. Second is Metcalfe's law, which means the value of networks grows as they increase in size. Third is Gilder's law, which means more data is transmitted with greater reliability and reduced cost by using computing technology.

These three driving forces, acting together, create the new reality your company must adapt to. For businesses in a new digital economy, your mission is not just to automate traditional manufacturing and administrative processes, but also to use digital technologies to learn about and solve customers' key problems and work with others within and across traditional company and industry boundaries.

As a business leader, you do not have to become an expert in the ins and outs of each technology, but you must develop the necessary insight to know how advanced technologies could challenge your business model, altering the initiates of revenue and shifting the sources of revenue. You need to shift your attention from thinking about how digital technologies *support* your current business to examining how they could also *shape* your future strategy and business models. To do so, you need to examine broadly and evaluate logically how to take advantage of technological developments and digital partnerships to create innovative new ways to create and capture business value.

WHY ISN'T DIGITALIZATION ON THE AGENDA?

To understand why you should focus on digitalization now, we must start with a broader, but very central question: "Why do successful companies die?" The answers reveal the reasons that limit companies from recognizing the opportunities and challenges posed by digitalization. There are four common "success traps" within traditional business models: the Competency Trap, the Ecosystem Trap, the Talent Trap, and the Metrics Trap.

The Competency Trap. Every company develops its core competencies, first by acquiring a set of physical assets and then by investing in people with the knowledge and skill to create its products and structure the organization with appropriate governance. These core competencies get progressively more refined and enhanced within the entire organization's strategic structure, and with a system aligned around them, defines how a company earns revenue and profit, and differentiates itself from competitors. These areas of expertise are the reason customers pay a premium price for particular products or services. Over time, these competencies become difficult for competitors to imitate, and these competencies are then considered core to the company's future performance. Business leaders stay with the current model because it works; there is little chance to question when—or whether—consumers might not value these competencies, and profit will fall. For example, let's look at Nokia. The company thought that its feature phones with short message service (SMS) and global reach could guarantee its future success. It was a leader in its respective domain. Then Apple introduced the iPhone, and Nokia's phone was rendered obsolete. The company was simply trapped in its historical competencies. More companies will become similarly trapped, especially during periods of digital transformation.

Here we will think beyond your current competencies and position yourself—and as a result, your company—for relevance in the digital future.

The Ecosystem Trap. Every company develops networks of relationships with supply chain partners, technology vendors, key marketing partner, research and development, and innovative companies. These relationships are built over time through trust and complex negotiations, and many companies develop specific organizational processes to maximize these relationships and reinforce the core competencies. The result is such: companies stay with the same suppliers and partners because they work. There is little time to question whether other relationships might bring better value, especially when they could conflict or challenge existing arrangements. In digital-era companies, these relationships are key: you need to rely on your traditional competitors as well as technological startups and large digital companies. These networks of companies, called ecosystems, both cooperate and compete with each other in new ways different than traditional business are used to collaborating. Take Microsoft for example. Microsoft dominated the personal computing ecosystems in the 1990s. The company developed strong relationships with hardware manufacturers such as Samsung, HTC, Sony and Toshiba to put, as Bill Gates famously said, "a computer on every desk and in every home." Microsoft focused on fine-tuning its ecosystem to keep up with newer versions of its software running on ever-faster computers powered by more and more powerful Intel chips. Then Apple came out with its smartphone, essentially a handheld computer that worked nearly everywhere. Microsoft did not recognize the value of a mobile operating system ecosystem, and missed an opportunity on mobile.

Here we will discuss the structure, and how to nurture ecosystems that cut across industry boundaries, especially as your industry's traditional disciplines collide with digital technologies.

The Talent Trap. Every company strives to acquire, nurture, and manage the best human talent, and to capture value from its expertise. It's a complex business, especially striking the balance between a deep expertise in specific domains while also being excellent in creating differentiation and adapting to changing contexts. When a company hires staff for existing roles instead of identifying the profile of talent that might be needed in the digital future, there is little attention paid to nurturing digital business literacy and onboarding staff with the vision and technical skill to implement new structural processes that enable digital transformation. Motorola, which invented the mobile phone, is a good case study. The company missed the smartphone innovation even as it worked with Apple to introduce iTunes to phones. According to a survey conducted by McKinsey & Company in 2015, the "most common hurdle to meeting digital priorities, executives say, is insufficient talent or leadership."

Here we will review the topic of how human talent can work with powerful machines to create new ways of organizing that reflect the digital economy where data and analytics at all levels become central and crucial.

The Metrics Trap. Successful organizations are particularly driven by metrics, most often quantitative measures of efficiency, quality, cost, and profit margins. Even performance is measured against a set of targets and individual managers and team and organizational units are driven to meet specific targets. Most metrics focus on short-term performance such as market share, sales per unit, and customer profitability. In the absence of long-term thinking about digital transformation and innovation, they reinforce a short-term focus and incremental changes in how and where scarce resources are allocated. When market share is the paramount goal, mergers and acquisitions favor familiar companies in conventionally defined industry boundaries rather than digital companies that might bring much-needed newer capabilities. If the automotive companies focused less on the number of vehicles sold and more on the number of miles that people traveled, it would be a new perspective to design business.

Here, I will show you how to use metrics that are subtle and directionally right so that you can successfully experiment with new digital technologies as part of transforming business.

These traps are not unique to digitalization, but one or more of these success traps very likely applies to your company and may be preventing you from taking the necessary steps towards digital transformation in your business. Recognizing the situation is an important first step. Rewriting your company rulebook is the second step toward reinventing your organization's relevance in the future. The digital modeling and its agenda will help you to develop your perspective on digitalization, overcome the success traps, and adapt your business for this new digital reality.

NEVERTHELESS, YOUR BUSINESS IS BECOMING DIGITAL

Now let us ask the question, "Is your business becoming digital?" Chances are you might say YES. The question is not about whether you have already implemented a lot of digital changes in your own company, it's more about whether your business is influenced and affected by different digital technologies already in use or will be in the near future. Here are six kinds of DIGITAL Influences you are definitely facing:

Big Data Analytics & Artificial Intelligence: affect your business process and how you make key decisions

Social Web: shapes your customers' action, interaction and consumption

Mobile Apps & Cloud Computing: are essential to how you deliver your services to individual consumers and enterprise customers

The Internet of Things (IOT): links all of your products through sensors and software to the broader machine web and the cloud

Robotics, Drones, 3D-Printing: are key drivers of the evolution of your supply chain

Cognitive Computing Algorithms: influence how you think of reinventing your business for the future

Count how many of the above characteristics apply to you, and it is certain that many more will apply in the very near future. Sooner or later, you, along with most companies, will find digitalization at the core of everything your business does. Let's suppose that you said no to all of the above characteristics. Do you have competitors for whom the answer is yes? Why do they see things so differently from you? How might that give them an advantage over you? And wouldn't you rather have that edge?

It's easy to see Google, Apple, and IBM as digital companies because they deliver digital products and services. But so are BMW and L'Oréal, which are reimagining their products as part of digital ecosystems in transportation and fashion respectively. It's straightforward to say that Tesla and Airbnb are digital companies, as their products and services are enabled through digital devices and interactions. But by the end of this book, you will feel comfortable looking at companies such as General Electric (GE) and Whirlpool not as industrial companies but as digital companies too. You will also begin to accept that pharmaceutical companies such as Merck, Pfizer and Johnson & Johnson are delivering greater value to patient health and wellness as they enhance their traditional competencies in medical sciences with data on how their drugs interact with individual patients. So, every industry faces its own digital future, and by extension, every company in it does as well.

THE KEY IS EMBRACING CHANGE WITH AN OPEN MIND…

Today we are at an inflection point with business scenarios in which old meets new. Old definitions of industry competition and organization do not make much sense, yet we do not have novel ways to define and demarcate pocket value created by digital technology. Old rules of strategy do not appear useful, yet we do not have new rules of value creation and retention.

We see companies born digitally in the postindustrial age emerge with management principles and practices that are very different from the companies born before them. This book provides the Digital Modeling as a management agenda, to help you understand the forces that are likely to influence the landscape you will encounter in a new digital economic world. We discuss this digital transformation mostly from the viewpoint of a traditional incumbent in an industrial age. You know that you have to transform, yet you do not know the tried-and-tested model, how to work on the practices, and when to move to next steps.

In part one, **Going Digital**, the book will first clarify the definition of digital transformation for the consensus of discussion, following by explanations of the changes in the digital economic era, and the specific modeling of digital transformation, as well as the expansion on the three stages in more detail so that you will begin to see where your industry and company are located and understand the challenge and opportunities to achieve digital transformation.

In part two, **Being Digital**, the book will discuss three winning moves, starting with how companies transform from platform to ecosystems for extension the business scope and scale while achieve business growth, how to transform from competition to cooperation, and co-create new business value, and how to design your organization to transform talent to intelligence with the powerful new intersection between humans and machines.

In part three, **Leading Digital**, the book will explore down-to-earth strategies and tactics of executives for a dynamic digital transformation, providing new perspectives for companies—including traditional incumbents, digital giants, and tech entrepreneurs—to be leaders in the digital economic space through the advantage of disruptive technologies, acquiring and applying digital marketing approaches, and engaging and maintaining consumer's lifetime relationship, as well as taking new actions with a digital mindset and value proposition, to lead in this new era.

Transformation is not easy, but it is inevitable, and it's necessary for companies to move from struggling, to surviving, to thriving. Do you feel confident about your company's continued success? Do you see any signs of success traps? Does your organization have the mindset and the skill set to deal with digitalization? No matter where you are within a digital transformation, you have a systematic guide to understand the process and its rules of digital modeling, helping you make decisions not only in the short term, but also for the long-term success of your company.

Time to open your mind wide and bravely embrace the changes. Let's start the journey of digital transformation in a new era.

PART ONE

GOING DIGITAL

CHAPTER 1
THE CHANGES IN A DIGITAL ECONOMY ERA

FIRST, THE DEFINITION OF DIGITAL TRANSFORMATION

Digital is one of the most frequently used business terms today, and also one of the least well-defined. Throughout the book, we will crystallize a *digital* concept that will guide us in the business of digital transformation.

The objective of digital transformation is to improve business performance and achieve sustainable growth. Organizational changes through the application of digital technologies and inventive business models are used to enhance productiveness and reduce cost. One or more digital technologies must employ a significant influence. Digital business transformation requires organizational changes, which include business strategy, operating process, as well as human resources.

Connectivity enables the convergence of multiple technology innovations. Naturally these innovations evolve over time, but the most relevant technology innovations today include:

- BIG DATA ANALYTICS
- ARTIFICIAL INTELLIGENCE & MACHINE LEARNING
- CLOUD COMPUTING
- SOCIAL MEDIA & OTHER COLLABORATIVE APPS
- MOBILITY SOLUTION & LOCATION-BASED SERVICE (LBS)
- THE INTERNET OF THINGS & CONNECTED DEVICES
- ROBOTICS, DRONES, 3D-PRINTING

- COGNITIVE COMPUTING ALGORITHMS
- VIRTUAL REALITY

Digital business transformation involves digital technology; meanwhile, it is quite beyond technology itself. The mission of digital business transformation aims at GOING DIGITAL, which means to change the mindset, and to build up an alternative strategic vision, organizational structure, operational system, and performance measurement of digitalized business.

THE THREE DIMENSIONS OF DIGITAL BUSINESS

In the Industrial Age, companies expanded their scale by increasing sales of their products and services, which resulted in a higher market share. And this scale expansion was linear based on the company's ability to access the physical, human, and financial capital necessary for growth. Think of Coca-Cola expanding globally and McDonald's opening stores across the United States and around the world. Some companies expanded their scope by extending their existing product lines and introducing new related ones. Think about Procter & Gamble's steady expansion in household and personal-care brands, organically and through acquiring other companies, from Ivory body soap to Pampers disposable diapers to Gillette razor blades. Such scope extensions were rather gradual, often requiring substantial financial and human capital. This steady and progressive expansion in scale and scope was a successful business strategy.

Digital companies show patterns of scale expansion and scope extension at a rapid speed that is quite different from that of the industrial era. Instead of linear rates of change, digital companies are showing mastery over non-linear, exponential expansion in scale and scope. In this scenario, they start to influence industries with new capabilities that take advantage of digital technologies. So, you need to position your business on the interconnection of scale, scope, and speed.

SCALE. In 1999, Google handled one billion search queries; by 2012, that number was 1.2 trillion in a single year, and up to two trillion searches annually in 2014. The company was not in the smartphone business in 2008, but by the end of 2015, more than 1.5 billion devices were running Google-powered

Android operating system software. That's Google's non-linear, exponential scale with search and mobile industries.

When making comparisons between digital business and their traditional counterparts, the distinctions are significant. Uber, which had a small set of drivers in 2011, had more than 300,000 drivers by the end of 2015, doubling from 150,000 just a year earlier. On December 30, 2015, it recorded its one billionth ride, and six months later it had reached the two billion ride mark. By early 2016, it operated in more than four hundred cities in 70 countries. By comparison, yellow taxis in New York City, which is considered a popular ride-hailing city, recorded about 175 million rides in 2014 according to the Taxicab Fact Book. Airbnb, the online accommodation booking service that started in 2008, had 50,000 listings in 2011, grew tenfold to 550,000 listings in 2014, and increased roughly fourfold to two million in more than 190 countries by early 2016. Meanwhile, the long-standing Marriott Hotels had about 760,000 rooms available worldwide in 2015. Amazon, the largest online retailer and Walmart's main competitor, recorded 304 million active customers by the end of 2015, compared to Walmart's 260 million customers in 2015. And, after just twenty years of existence, most of Amazon e-commerce (EC) competitors have withered away or become smaller niche players. Although several large retailers coexisted in the industrial age based on locational advantage and differentiated merchandise, digital-era retail seems to favor one or two large players alongside countless specialist players, a phenomenon that has come to be known as the long tail.

In the 19th and 20th centuries, railroads and telegraph lowered the costs of transportation and communication, and the successful firms were the ones that built and maintained the organizational capabilities necessary to exploit economies of scale. They invested in the capital equipment necessary for high-volume production with local and global networks of marketing and distribution. And they formalized organizational structures and management systems that allowed them to invest to take advantage of economies of scale. In the words of historian Alfred Chandler, "The modern industrial corporation of the twentieth century exploited economies of scale because of the three-pronged investment in production, distribution, and management."

McDonald's is one of the most typical examples of the linear growth of the traditional leading companies. McDonald's reached the one million people served mark in 1955, the one billion people mark in 1963, and in 1994, 100 billion. And then the restaurant signs just simply said "Billions and Billions Served" and left it at that. Why? Because McDonald's did not keep track of how many individuals consumed their hamburgers. They can only count the number of hamburger patties shipped to all their locations.

That's the difference now. Digital-era businesses such as Google, Uber, Airbnb, and Amazon, have amassed detailed data on their operations. Google's database of our search queries is now in the tens of trillions, and the same holds true for its mobile platform. Uber collects data on its one billion-plus rides and uses it to fine-tune its operations in ways that the taxi companies haven't achieved. Airbnb knows where, when, and how long we stay in ways that the hotel chains do not and cannot. Amazon knows our buying habits in richer detail than Walmart ever has. Digital business knows consumer preferences and behaviors in ways that traditional companies never did or could: they simply were not designed to collect, process, analyze, and interpret such data.

Therefore, if you are still operating on the assumption that scale means the number of products manufactured or sold, and that selling more units relative to your direct competitors means a higher market share and subsequently a lower per-unit cost and higher profitability, you may be at a scale disadvantage in the digital world.

SCOPE. How did Amazon go from being an e-commerce bookseller to a soaring giant on the cloud in just twenty years? How did Google parlay its supremacy in search into leading the mobile and media web, and into automobiles and health care within a single decade? How did Apple go from being one of many companies selling personal computers in 2001 to dominating the music and telecommunications industries by 2011?

In the Industrial Age, companies traditionally expanded their scope incrementally and relatively methodically by testing and then extending their core competencies in new geographies or market segments, or by gradually adding products that served their core offering. For example, meat-packing firms took advantage of by-products in their production processes to make leather, soaps, and fertilizers. And Honda used its core engine technology to offer motorcycles, automobiles, lawn mowers, and aircraft engines. By the end of the twentieth century, General Electric was selling not only the home appliances for which it was known, but also unrelated products and services such as aircraft engines, entertainment, and financial services. At that time, most companies that diversified too far from their core markets, such as United Technologies, which was funded as an aircraft manufacturer, and ITT corporation, which was originally a telecommunication company, were brutally punished by the stock market, and they returned to their core competencies and diverted non-core business and processes.

In contrast, digital-era companies rely on their core competency of data analytics to predict with a high degree of accuracy about what their consumers want. With machine learning and artificial intelligence, Google, Uber, Airbnb, and Amazon also can take huge volumes of data and analyze it to expand their scope with new products and markets, even in unrelated industries. And because they can track when, where, and how consumers are reacting, it is easy to adjust the strategic execution quickly; the digital businesses can minimize risks and capitalize on successes to fuel exponential growth.

The bottom line is that you are still operating on the assumption of keeping scope means extending your reach only within your own or adjacent industries. With products or services related solely to your historical core competency, you may be at a scope disadvantage, and if you think that you are only vulnerable to competition from leaders in adjacent industries, you are looking too narrowly at the landscape.

SPEED. One of the famous mottos from Meta (Facebook) CEO Mark Zuckerberg is: "Move fast and break things…unless you are breaking things, you are not moving fast enough." That is exactly the idea of speed in the digital world. It's not about being reckless; it's about continuous improvement and iteration, a culture that Zuckerberg calls the Hacker Way—because hackers believe that something can always be better and that nothing is ever complete. Using that same hypothesis, Google develops products in the open, and features them daily or weekly, and closely observes how customers use them. This instant feedback makes customers trusted co-developers.

In the Industrial Age, companies hurried to lock up physical assets such as land and machinery, as well as access to production and transportation. Traditionally, speed referred to the time it took a company to act to changes in the specific industry and relative to other competitors within that industry. Writing in the late 1980s, George Stalk at Boston Consulting Group argued, "companies that meet the needs of their customers faster than competitors grow faster and are more profitable than others in their industries. We argued that time could be the next decade's most powerful competitive weapon and management tool for companies." Viewed this way, your speed allowed you to reap first-mover advantage relative to others competing against you within the accepted industry definitions. In other words, your ability to be faster in the market hinged on your organization's own clock speed in areas such as product design and development, manufacturing and supply chain synchronization, and so on. It also depended on your information technology department's ability to speed up the back-office processes, which is often operating on antiquated systems and legacy infrastructure, to support the development of new products. The slowest part of the interlinked processes defined your speed. As long as your competitors were in a similar state, this did not prove ruinous.

Now, the digital players are dictating the pace of customer service with new services that are enabled and delivered via the cloud and through apps on the mobile phone. You not only need to speed up the back-office processes to compete against your traditional competitors but you also need to calibrate the speed of your delivery to the benchmark set by companies born in the digital era. If you are still operating on the assumption that speed means being the first to move into a new market, rather than the fastest to capitalize on the opportunities, you may be at a speed disadvantage.

THE TWO ADVANTAGES OF DIGITAL BUSINESS

In Industrial Age companies, scale, scope and speed acted independently. The scale decisions were handled within individual business units, which first sought to become efficient in production or distribution at the minimum viable scale before expanding based on the available resources, organic growth, and acquisitions. The scope decisions concerning corporate strategy often involved mergers, acquisitions, and joint ventures, in addition to signify realigning the resources of existing businesses. The speed often reflected the pace towards market and defined a company as either slow or fast relative to other competitors within specific industries. As an incumbent in a traditional industry, you already know how to tap into the advantages of scale, scope, and speed within your industry. You may have developed an advantage in one or more of these dimensions compared to your traditional competitors.

As your industry goes digital, progressively in some cases and rapidly in others, you need to look at these three dimensions of your business as being interconnected. Scale and scope still define your company's strategic ambition, but you must address this question: What set of businesses should we operate and at what scale? However, scale at speed did not create a first-move advantage but a fast-mover advantage, which may currently be limited by your company's internal organizational processes and systems—if they cannot recognize and respond to the shifts as quickly as some of the newer companies. Changing scope at speed also reflects a fast-mover advantage, where the advantage may not necessarily lie in launching product but in tapping into scarce critical resources such as unique interconnected data, patents, talents, or research and development projects, often executed with others.

How well you stack up against not only other incumbents, who themselves are transforming, but also against newer companies that aim to disrupt and transform your industry may well define your ability to compete and win in the digital realm. Those companies that take maximum advantage of scale, scope, and speed together are able to gain significant advantage in the digital business world. First, with data and analytics and connectivity, you can now extend your footprint beyond your firm's core boundaries and tap into extended ecosystems. Second, through sensors, software, and connectivity, you now have the capability to collect data, process information, and learn in ways that would have been difficult—if not impossible—in the industrial world.

The Ecosystem Advantage. Whereas scale advantage arose in the industrial world from assets that a single firm controlled and the units that it produced, in the digital world, scale advantage comes from being part of an ecosystem that includes key partners that play complementary roles. Ford and GM's scale depends on the number of cars they produce, but Uber's scale is defined by the number of cars it has in its network on a global basis as well as locally in every one of the 400-plus cities in which it operates. Whereas Nokia's scale depends on the number of feature phones it manufactures and sells globally, Google's scale advantage, as the architecture of the Android mobile operating system, depends on the number of devices produced by its hardware partners in the ecosystem and the number of software apps written by the developers for its operating system. In the Industrial Age, scale is the result of what a firm does by itself using the assets that it controls and the units it produces. In the digital world, scale is the result of what it may produce by itself, plus what it can achieve with its partners in the ecosystem. Tap into the scale advantage conferred by your ecosystems.

Just like scale, scope advantage in the digital world comes through being part of an ecosystem. In the Industrial Age, the relationship between a company's core area and its adjacencies had to be pretty close for customers to accept the link. On the contrary, in digital, data as a core area is infinitely malleable so companies that collect data can more easily apply it across a wide range of platforms. For example, in mobile platforms, with their core mobile software-Apple's IOS and Google's Android—digital giants can logically extend their scope with different apps. Payment apps, such as Apple Pay and Android Pay, supported by merchants and global retail banks create an ecosystem that allows Apple and Google to move into the seemingly unrelated area of retail finance. But they do so for different reasons. Apple does it to enhance the use of its phone and watch, but explicitly is not using the information on such transactions, and Google is does it to better target its advertisements by using that information.

In contrast to the industrial world, where a single company could gain an advantage by being the first one into a new market, in the digital world, everyone in an ecosystem has to move at more or less the same speed. Since not all the competencies lie inside your firms, you have to use a relay on the ecosystems. It's like being on a relay team: one super-fast runner is not going to win the race for the group, though one super-slow runner, like one super-slow company, could lose the battle for the entire ecosystem. In other words, your critical skills and capability to stay up to speed could well be the deciding factor. Sony PlayStation has succeeded over the past decade because it has mobilized its game development partners with the pace of successive console developments.

The Learning Advantage An important characteristic of scale-scope-speed is learning from products and services in use and adapting their characteristics to the specific needs of individuals. So how do we think about collecting data? Whereas companies in the industrial world collected data about a few attributes, mostly focused on operational efficiency, and analyzed this aggregated data over time, digital companies are constantly recording data with detailed attributes and analyzing it using new tools to discern patterns of preference and adjust their strategies. To determine how many burgers it sold, McDonald's counted the total number of hamburger patties shipped to its locations. In contrast, Starbucks uses its apps and loyalty programs to understand not only how much coffee it sells but when and where its customers buy their coffee, how they prefer it, how much they spend per transaction, and so on.

Products behave differently under various conditions and no amount of testing in the lab is enough to understand the particular behavior in the real-world— whether they are tractors on the field, aircraft engines in flight, or washing machines at home. Now companies monitor their products in so many different locations at scale, and in near real time, that they have more opportunities than ever before to learn about them, modify them, and even correct mistakes before the impact is felt too widely. They learn from products at such a scale to gather early warning signals.

Companies in the industrial era expanded their scope by branching out to related products or markets, and they made these decisions based on pre-established patterns, followed by others, and based on analyzing data from market research and other rough data. In the digital age, companies can actually predict areas of inefficiency by using analytic software and expand into seemingly unrelated areas. For example, GE, after borrowing from the playbooks of Apple, Google, Microsoft, and others, is now on a new mission to use software, apps, and data plus analytics in four industries: buildings, power, industrial transportation, and health care. The company's Predix platform, with analytics as the foundation, allows it to predict areas of major inefficiencies within and across diverse industrials and solve them better than anyone else, including their own customers. Furthermore, we can now not only collect data on our own products, but also see how products from different companies operate together to solve customer problems. For example, in a health care setting, firms can monitor how their device or medication interacts with other treatments across a wide spectrum of different patients. Bearing in mind proper safeguards of privacy and security, all the companies contributing products can learn from the data and tailor their products to individual patients, specific treatment plans, and any number of other variables. Similarly, companies such as Amazon, Google, and Facebook have access to customer data that could be mined for an advantage, learning from customers that use complementary products to proactively improve key features.

Industrial Age companies spent a lot of time before starting experiments to make sure that the goals were well specified and the mandates well established. Digital age companies start projects on the backs of passionate people who try, fail or succeed, learn and adapt. They fail fast and pivot, which simply means they learn fast with data, adapt their prototypes, and reflect on customer feedback. They pivot along different dimensions, such as customer segments, channels, revenue streams, partnerships, and value propositions. Since every interaction is an opportunity to collect data about the products and systems in use, they move fast to embrace new ideas not because they are slavish, but just to learn at a deeper level. The ultimate advantage at the nexus of scale-scope-speed, then, is reflected in learning through experimentation and taking advantage of the greater scale and scope of your ecosystems. For example, Netflix used machine learning, analytics, and A/B testing to create its personalized video recommendations. Doing so at speed—understanding the validity of how your assumptions operate and iterating fast based on the results—allows you to refine your working hypothesis in key areas. Learn from experimentation through data and analytics.

MASTERING YOUR EXPONENTIAL GROWTH PATH

Learning from ecosystems is continuous. As ecosystems help you scale further, you gain more opportunities to acquire knowledge. As ecosystems help you expand the scope of your business footprint, your learning opportunities expand as well. And as you extend your scale and scope at a faster speed, you increase your learning opportunities further. So, scale, scope, and speed are mutually reinforcing. What emerges at the nexus of scale-scope-speed is a new focus on a non-linear, exponential growth path, and your ability to master these shifts as your industry digitizes and evolves exponentially is an important new strategic requirement.

To understand this non-linearity, let's look at an example, Ray Kurzweil, author and resident futurist at Alphabet, believes in a "law of accelerating returns" arising from the exponential increase in the power and functionality of personal computers and smartphones. He traces the ever-quickening evolution over the past 100-plus years from "the mechanical calculating devices used in the United States in 1890, to the Turing relay-based machine that cracked the Nazi enigma code in World War II, to the vacuum tube computer in 1952, to the transistor-based machines used in the first space launches, to the integrated-circuit-based personal computer over the last 110 years. Looking ahead, this exponential increase in the functionality of computing power will extend to other areas such as devices connected to the Internet of Things, wearable computing devices embedded into clothing and footwear, health care devices, drones, 3-D printers, robots, and automobiles. As the number of such powerful network-enabled devices increases to 50 billion or more over the next decade, managing the exponential shifts in digital business will become the top priority. Non-linearity in technical features and performance improvements may be obvious to the technologists on your team, but you should recognize and respond to the opportunities and threats in this new business landscape of cross-industry ecosystems and make efforts on extended social and professional networks.

CHAPTER 2
THE DIGITALIZATION AGENDA

THE FRAMEWORK OF DIGITAL TRANSFORMATION

In the digital era, business games have new rules. Here is a framework that allows you to see how three types of players use a variety of digital technologies to shape the future of your industry, and influence your company's strategic actions and response over three distinct phases of digital transformation. In such a new digital economic world, all the players are adapting to design effective business agendas for the future using three winning moves. Picture Digital Modeling, as a control agenda of digital transformation, with nine charts of three players along with the other axis of three phases.

THE DIGITAL MODELING		3 PHASES		
		EXPERIMENTATION AT THE EDGE	COLLISION AT THE CORE	REINVENTION AT THE ROOT
3 PLAYERS	DIGITAL GIANT			
	INDUSTRY INCUMBENT			
	TECH STARTUP			

Remember that Digital Modeling is a management framework. It is not a technical or tactical one. It invites discussion and debate, raises options and investment trade-offs, and calls for strategic solution for actions. It is a call for you to give up some successful, but now outdated, past practices, and to embrace new rules for continual experimentation with new approaches to suit your needs. Three main characteristics set the digital agenda apart from other approaches.

It is not a framework of technologies. The forces of digitalization are not simply a set of technologies. The Digital Modeling recognizes that you are developing your strategy against the broader landscape of decisions and sequenced actions taken by three sets of players, and that these players, not the power and functionality of technologies, drive the transformation. It takes into account that companies in different roles and phases are all embracing, experimenting with, and exploiting digital technology to craft new business logics that give them some advantages, but they are absorbing and assimilating it differently to create new capabilities, establish new relationships, and seek differentiated drivers of value.

The digital agenda captures the core actions by different types of players rather than focusing on the devices that drive them. In other words, it's more focused on business influence rather than technological influence.

It is not static. Digital strategy is not a set of specific actions carried out in a particular order at a specific stage of a predefined life cycle, or with new technologies. Digital Modeling recognizes that companies and industries are constantly moving through three phases of transformation that evolve quickly; therefore, it does not try to impose a one-size-fits-all solution. It takes into account that the players, technologies, and moves are continuously changing and creating new conditions.

The digital agenda brings out the dynamics—the actions, reactions, and the follow-on decisions—to ensure that the players are continually taking advantage of developments with different technologies to evolve business for the digital future.

It is not one-dimensional. Success depends on being able to focus on three players and three phases simultaneously. Isolating any one variable is not enough, and it's this idea that is the key to Digital Modeling. Once you realize that your thinking must transcend traditional boundaries, that you must examine not only your own actions but those of others in the game, then you will see what is happening in other settings, better understand how to set the rules of the game, and be in a position to create the strategic alliances and big moves necessary to exploit new opportunities and win big.

The digital agenda sets out the multidimensionality through the nine screens, such that your actions in every phase, which are deeply interconnected and mutually reinforcing, invite reactions and responses from the three sets of players.

Start thinking holistically about how digitalization could drive the design of your business strategies and the structure of your organization. Consider how the different forms and functionality provided by digital technologies could provide better product and service value for customers. Recognize that the real payoffs happen when you can access and analyze data across functions and across companies, and then make changes to the products, processes, and services that connect all of them.

This transformation is about systematically recognizing how digital technologies—the ones that we have seen, the ones that are on the horizon, and the ones that may still be in development in research labs and universities—begin to change the scale, scope, and speed of your business. They start to shape your strategy in unprecedented ways. You need to look at the entire business system, and that means identifying and analyzing all of the players that intersect with you in this digital strategy game, represented across the nine screens.

THE THREE TYPES OF INTERACTIVE PLAYERS

In the digital world, you will find yourself competing against a broader set of players than the ones you may be familiar with in the industrial arena.

PLAYER ONE: INDUSTRY INCUMBENTS You know your current competitors, what I call industry incumbents. Chances are good that you have scanned, analyzed, and understood how these historical, traditional competitors in your industry respond to market shifts. You recognize them well enough to anticipate their likely actions and you have a range of responses to them. However, you are likely to be positioned in networks comprised of not only traditional competitors you know well, but also newer ones you don't know.

PLAYER TWO: TECH ENTREPRENEURS Ambitious, upstart tech entrepreneurs have a brazen view on how they can disrupt and reorder the business world. For example, PayPal, in financial services, Tesla in automotive, and Uber, in the ride-sharing sector fall into this category. These companies and others are born digital with a disregard for management rules from the industrial age. They believe in crafting business models that promise to deliver unparalleled value to customers by using the power of digital technologies. Algorithms and automation guide their thinking, data becomes their differential resource, and analytics their distinctive competency. They think beyond narrowly defined industry boundaries and evolve their business models by taking advantage of different digital technologies. Not all upstarts will succeed, but those few that do could turn out to be your formidable future competitors or trusted partners as you transform.

PLAYER THREE: DIGITAL GIANTS The third category of players is digital giant. These companies, including Alphabet (Google), Amazon, Apple, Facebook (Meta), IBM, and Microsoft, have progressively extended their influence beyond their traditional industry into yours. They were yesterday's tech entrepreneurs that have now grown up and extended their scope into industries in which they previously only supplied technologies or managed back-office operations. Their core business model may still be to deliver digital products and services, but some of these giants have increasingly partnered with industry incumbents to help those companies transform their business models for the digital age. The end result for such giants has been more vertical and horizontal integration into certain industries including yours.

You and your fellow industry incumbents will be in ecosystems that include tech entrepreneurs and digital giants. What you will discover is that in such ecosystems, not all relationships are competitive and adversarial. Some are cooperative and others are competitive. In this new business landscape, relationships that are cooperative in one time period may become competitive in another time period, and vice versa. And some of these relationships will become simultaneously cooperative and competitive, which is called "coopetitive". So, the digital business ecosystem is a fast-changing playground. You need to understand the different viewpoints and capabilities of each of these players in order to know how best to interconnect your company with them.

THE THREE STAGES OF DIGITALIZATION

Industries change. They grow, they shrink, they transform. That is not new. In the digital age, these transformations occur much more rapidly, and they are no longer linear or chronological. Incumbents must simultaneously manage their own actions and interactions with all three sets of players through three phases of evolution that exist at every point in time.

PHASE ONE: EXPERIMENTATION AT THE EDGE There is the primary phase, during which experiments with digitalization happen and evolve. As one digital idea matures, another experiment emerges, such that digital business experiments are happening all the time. This phase is called experimentation at the edge. This is when lots of ideas, some incredible sounding ones, and others more realistic and potentially valuable, going from sketches on napkins and slide decks to prototypes, pilots, and products.

During this phase, you could already be focused on the experiments that you are engaged in to adapt your business model. That is necessary, and you should focus on them. But the threat of this phase is looking at the landscape of experimentation undertaken by a wide range of firms—even those beyond your immediate industry boundaries—to make sense of their potential implications to your industry's ways of working. One prominent area with digitalization is the shift from individual products to interdependent products linked via platforms. In essence, a platform is a base on which others can build a business. Think of a computer operating system, like Apple's IOS or Google's Android. It is multi-sided in that it connects many different types of companies transacting with many different types of customers and gives them a standardized way to get paid or exchange value of different kinds.

Does this trend complement and enhance your current business model or could it fundamentally transform and disrupt it? When do platforms arise and how could they affect you? Even if platforms have not emerged in your specific setting, you have to understand the conditions that could give rise to new platforms, inducing competition and transformation. Hotel industry executives were not expecting platforms to upend their business logic until Airbnb showed the way. Beyond platforms, your requirement in this phase is to make sense of the signals at the periphery of your current industry, and pre-established business models, that could challenge the fundamental assumptions that underlie how your business earns revenue and makes profits.

Making sense of experiments at the edge involves deep thinking. The deeper thinking of this experiment at the edge should raise follow-up questions such as: What else is needed technically for Apple to launch its own mobile phone? If music is one application that is readily suited for mobile phones, what other applications are possible—and probable—with future developments in software and user interface? Such questions must lead to subsequently richer investing and analysis with the right mix of management and technology competencies to connect the dots and compute the probabilities of plausible innovations.

Take this same approach to deeply interpreting the possible outcomes when you look at other areas, such as drones from Amazon, virtual reality headsets from Microsoft and Facebook, chatbots from Apple, and so on. Think through what drones could do beyond providing logistical support to farming, mining, and disaster assistance. Imagine virtual reality beyond gaming in areas such as education and training in specialized areas of health care. Consider the second order privacy and identity implications of chatbots such as Apple Siri. When experiments mature from labs and research organizations to the mainstream, they become ready for prime time, to serve as the foundation of newer, innovative business models explored by any of the three types of players involved in the digital business frame. That shift gets us to phase two.

PHASE TWO: COLLISION AT THE CORE In phase two of transformation, ideas have evolved from prototypes to business options. Experiments have shown the difference between possibilities. Whether it works and what are the probabilities? How do we make it work so it's profitable? This phase sees genuine tension between established ways perfected in the industrial age and new digital ways experimented with just recently. This phase is labeled as collision at the core, where digital rules challenge traditional industry practices and pre-established rules of engagement.

This conflict seems to happen gradually in some industries and much more rapidly in other settings. Even within a single industry, some companies may face the intensity differently from others based on their unique strategic differentiation.

And the pace of change can accelerate suddenly.

Collision at the core occurs because digital technologies make an impact in two ways. One is that digital products and services challenge traditional products and services. Traditional analog wristwatches vie against smart-watches and digital wristbands. Conventional standalone refrigerators from GE, Bosch, and LG Electronics compete against those connected to the Internet of Things network by Samsung and others. As more products become digital, this collision intensifies, and competitive interactions between traditional industries, the digital giants, and ambitious entrepreneurs are exaggerated.

The other way is that newer organizational models are based on principles of computer science, such as automation, algorithms, and analytics with software models, rather than mechanical engineering rules that gave credence to scientific management principles. The result is continuous experimentation, dynamic coordination with partners in ecosystems, and machine learning at speed with powerful machines and smart humans working together rather than standardization, specialization and value-chain optimization. In other words, the collision between industrial-age companies and digital-era companies is a clash of assumptions around value delivery and organizing logic. Those who survive this clash live to influence the new rules, which is the third phase.

PHASE THREE: REINVENTION AT THE ROOT In this time of evolution and reinvention, digital ways of thinking are no longer an afterthought. Instead, industry incumbents, tech entrepreneurs, and digital giants work together to solve core problems for consumers by using digital functionality. At the center of every value proposition, whether it is a product or a service, every offering is digital and every business is digital. And traditional distinctions such as B2B or B2C dissolve; every company is situated in a B2B2C network of interactions. The new focus becomes: Who takes the final response for interactions with the end consumer? Who designs the interface to interact with that consumer? Who gathers and analyzes the data and develops insights? This phase is reinvention at the root. Reinventing the rules requires a new mindset, because it is about earning the customer's trust in order to monitor these interactions in fine detail and guarding the privacy of such data. It really is about finding solutions to the pain points and fundamental thorny problems facing consumers, either individuals or businesses.

Reinvention at the root is about strengthening the intuition and judgment that have been valid modes to arrive at decisions in the past with tangible support data and analysis. Developing insights and applying them faster and better than competitors is a key to success in this phase, so agility and the ability to adapt quickly are essential. To be fast and nimble, companies use data and analytics but more importantly, shed the bureaucracy that has been characteristic of the industrial world.

With your new understanding of the two fundamental axes of the digital modeling, the players and the phases, imagine that you are situated in one of the nine charts on the control agenda. You know already that the game is not going to be played out on just that one screen, the moves and countermoves taking place in the other charts will influence your moves and countermoves. Thus, you should position yourself on the digital modeling. Then you can better understand your position relative to the other players.

POSITIONING YOURSELF ON THE DIGITALIZATION AGENDA

No matter what industry you are in or what service or product you offer, your company is located at the nexus of three players and three phases as shown in the following figure.

The digital modeling helps you understand exactly what your position is at this point, where you are situated among the nine charts on the control agenda. If you are a tech entrepreneur, your position is along the bottom three screens, if you are a digital giant, you are along the top three. If you are an industry incumbent, you are in the middle set. That is defined by the first criterion, the type of player you are.

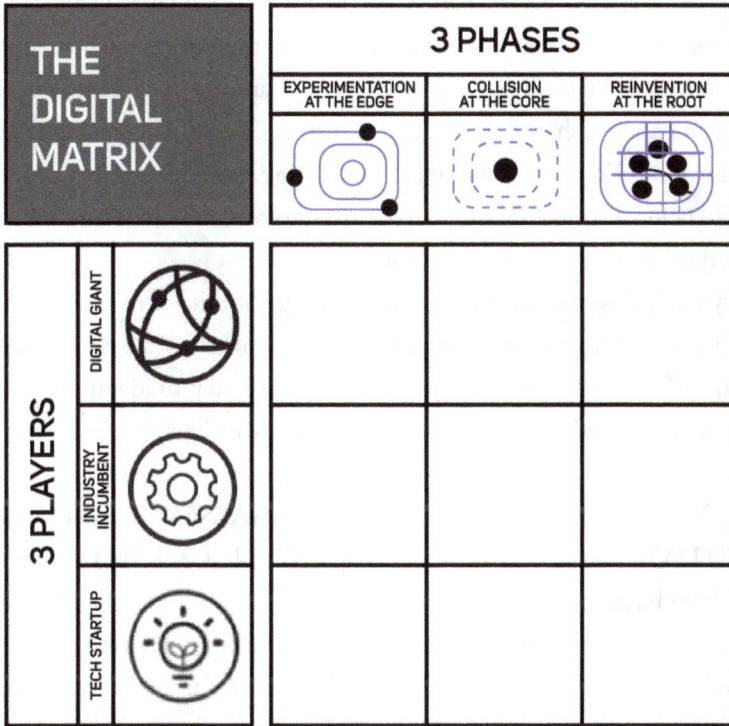

THE DIGITAL MATRIX		3 PHASES		
		EXPERIMENTATION AT THE EDGE	COLLISION AT THE CORE	REINVENTION AT THE ROOT
3 PLAYERS	DIGITAL GIANT			
	INDUSTRY INCUMBENT			
	TECH STARTUP			

IDENTIFY YOUR INDUSTRY ALONG THE THREE PHASES. Is your industry in the experimentation phase, where you can see digital possibilities within your own company and other industry incumbents, or where tech entrepreneurs or digital giants are dabbling with some interesting innovations, but your core business model has not changed? The global agriculture industry is in this phase: experiments with sensors, climate data, and tractors connected to the cloud have been happening in recent years, but the agriculture business model remains fundamentally unchanged. However, precision farming and digital agriculture are on the horizon.

Is your industry in the collision phase, where upstart entrepreneurs and digital giants are using powerful technology to challenge and disrupt the way you and your traditional competitors do business? The global automotive industry is at this stage, where the industry incumbents are being forced to rethink their business models by upstarts that redefine product architecture and use a transportation service as an alternative to car ownership.

Is your industry reaching the reinvention phase, where industry incumbents, tech entrepreneurs, and digital giants are all using powerful technology and new approaches to solve fundamental business problems? The media and entertainment industries are being fundamentally reshaped by companies such as Google, Amazon, and Netflix as well as Disney and Comcast, which are all monitoring customer preferences and using that data to tailor their offerings to individuals, fine-tune their products and services, and pivot quickly when changes are needed.

Find your industry's current position among the nine charts. If you are an industry incumbent, you can now more precisely situate yourself in one of the three charts of the middle band. But do take the next step to identify the relevant tech entrepreneurs and digital giants that are already or likely to be in your orbit.

ASSESS YOUR RELATIVE POSITION AGAINST YOUR DIRECT COMPETITORS AND OTHER INDUSTRY INCUMBENTS. Companies vary widely in terms of the importance they give to new trends, and digitalization is no exception. Are you leading relative to other industry incumbents? Have you recognized the certainty of digitalization and taken an early lead over your other industry incumbents by making acquisitions and entering into alliances to gain initial advantage? Do other incumbents look to you, as incumbents in the manufacturing industry look to Siemens or Bosch in Germany as benchmarks, to understand possible directions to follow with digital shifts?

Or have you, like many firms that think digitalization is not yet relevant as a company-wide strategic issue, allowed others to lead with digitalization? If you have consciously pushed digitalization to the back burner because of other pressing challenges, keep track of how others in your industry have invested in digital technologies. It is more important to understand the pattern of shifts in other industries that may provide pointers for you. Or perhaps you are in the middle of the pack, together with other incumbents that are paying minimal attention to digitalization where it's needed but nothing more.

Calibrating your position on the lead-parity-lag scale is a useful exercise. Ultimately, it does not matter whether you have already taken any steps relative to other traditional industry incumbents. What matters is that you are prepared to systematically accept digitalization as a transformative force in your industry going forward. Knowing your position on the nine-screen grid of the digital modeling is a good start. But it's just the beginning. No company is an isolated island in the digital world. Value creation and capture involve interlinking islands. So digital modeling recognizes that tech entrepreneurs and digital giants are potential allies, but also possible adversaries. The actions and interactions they take amongst themselves, with your industry incumbents, and with you, profoundly influence your ability to craft winning strategies. Take action. What matters is that you are prepared to systematically accept digitalization as a transformative force in your industry going forward.

CHAPTER 3
THE STAGES OF DIGITAL TRANSFORMATION

EXPERIMENTATION AT THE EDGE

Looking beyond the obvious, connecting the dots, and developing a compelling narrative to guide your future business decisions is a valuable reason to experiment and begin phase one of your company's digital transformation.

Scan widely but connect the dots. If you think back to the nine charts on your digital modeling agenda, the three charts in the first column represent the various experiments carried out by the three sets of players. These are the different initiatives from eccentric entrepreneurs, bold experiments from digital giants, and thoughtful extensions from incumbent companies in different industries. They are experiments unconstrained by past definitions of industry, or functions, or geography. You want to scan widely, but your objective is clear: to make sense of the experimentation frontier and guide your company's transformation. Your ability to connect the dots across a wide range of experiments by the three sets of players is key to knowing your business wins in the digital future.

You know that future business models lie at the nexus of seemingly disparate, diffused, and disconnected trends such as social media, block chain, artificial intelligence, and wearable technology. And you see where these trends might converge and create compelling new business rules. This is more than coming up with mobile apps to post web content to smartphones and tablets. It is more than creating Facebook pages for your brands or striving to imitate Uber or Airbnb in different industries. You are especially looking to discover those ideas that might have been dismissed as science fiction one day, like 3-D printed cars or modeled human organs, then have become very real and commercial successes the very next day. Understanding how, where, and when they could become real in prototypes and at scale is important.

You are scanning widely to develop a story—a compelling business narrative well supported by data and analysis—of how digitalization could shape the future of your industry. You are curious to see how disjoined technologies could interconnect to unleash new business functionality, just as mobile and social webs intersected over the past few years. You are interested in knowing how distinct technologies such as 3D printing, robotics, and drones emerged as driving forces to redefine the geographical distribution of high-value manufacturing activities. Your challenge is no less than developing a theory of digital business by laying out plausible scenarios from experiments occurring at the edge of your own industry and in other industries too.

To use a photography analogy, you need a powerful lens that combines a wide aperture and a long depth of field. Interestingly, all three sets of players may be using the same type of lens but looking for different objects and trends. That's what makes this phase challenging and powerful.

ENTREPRENEURS: LAUNCH AND EVOLVE THEIR BUSINESS. The story of Netflix, which disrupted the movie rental business, caused the bankruptcy of Blockbuster and other video rental chains. The more important chapter of the Netflix story involves the experimentation underway right now. The DVD-by-mail business, which was the core of Netflix when it started, is a tiny part of the company today. Through video streaming, the company has steadily transformed itself into a leading online television network that could replace the broadcast TV business model of the last century. Internet TV is on demand, personalized, and delivered from the cloud to any screen with online access. More than 75 million subscribers pay their fees in more than 190 countries, these subscribers enjoy 125 million–plus hours of streaming TV shows and movies.

Netflix has been a consummate experiment. Multiple trials over time have allowed the company to steadily fine tune business models for the digital media age. All of the experiments have been at the edge of the TV and entertainment industry, but the mainstream industry leaders have never taken them seriously, until perhaps now.

While most others in the movie rental business focused on top movie hits based on general lists, Netflix focused on developing recommendation engine software that world learn an individual's film preferences and use that information to suggest other appropriate titles from its catalog. At first, the engine relied on individual ratings and preferences from previously watched films to predict user ratings for other movies. Later, it pulled together data from similar profiles of viewing habits. To further ensure the superiority of the recommendation engine, the Netflix Prize opened a competition for anyone to improve the "Cinematch" filtering and recommendation algorithm. Then, Netflix experimented to add unlimited streaming to its DVD subscriptions, to understand how people were embracing online streaming. Today, Netflix is the undisputed leader in streaming video, and personalization is a big reason for that success. Each subscriber's homepage shows groups of videos arranged in rows with a title that conveys the meaningful connection between the videos in that group. Netflix becomes the leader at personalization, which is done by machines and at scale and speed. Meanwhile, Netflix settled on software to make streaming work across multiple devices-TV screens, set top boxes, TV sets, personal computers, tablets, phones, and others.

As you made sense of Netflix, you realized that the experiments undertaken by Netflix teams were not one-offs, they were sequenced to marry developments in technologies, with plausible options for superior customer value and different business models, that place digital characteristics such as algorithms and analytics as well as streaming and cloud as key anchors. Such sequenced experiments, refined over time, allowed Netflix to become a global powerhouse in streaming television.

INCUMBENTS: EXPERIMENTS TO COMPLEMENT OR CHALLENGE.

What digital experiments should incumbents track? The answer lies in two types of experiments at the edge as early stages: those in which experimentation complements your current business models and those in which it challenges them. The former gives you early indications of how you might proactively embrace those ideas, and the latter points to warning signs you should consider.

In 1987, Nike introduced a device called Monitor, which had a strap with sonar detectors that tied around a runner's waist and could calculate speed and announce it through the runner's headphones. This was the time of Sony's Walkman, and Nike was very much rooted in the business of sports footwear and apparel. Nike had always dabbled in digital technology and this was one of those experiments. The Monitor failed because it was way ahead of its time. Then Nike tried selling branded sports watches and heart rate monitors and they failed again because it was way ahead of its time.

However Nike persevered and persisted in its belief that digital features could complement their physical products, experimenting with designs that led to the release of a smart shoe in 2004. This was way before the emergence of a network of smart devices that we now call the Internet of Things (IOT), so the smart shoe was again, well ahead of its time. What made this experimentation distinctive and instructive was that it marked a joint collaboration between two companies, one in the traditional industry and another in the nascent digital sector. Nike worked with Apple to create Nike+, a running shoe that logically connected running with music and data collection. In this business scenario, Apple has its portable music device and the iTunes technology allowed Nike to seamlessly sync data at that time through a personal computer acting as the interface. Apple refined Nike's prototype sensor while Nike focused on the shoes and the interface for the web and the iPod. Essentially, it was the beginning of the quantification of the self-movement, in which we now use apps to measure how we live, work, and play.

Incumbent leaders are prepared to accept the possibility that their current business model and accompanying assumptions about impermeable boundaries among industries, narrow scope, functional specialization, and value chain, and thus, are out of date for the digital age. When that happens, they are prepared to experiment not just with refining or adjusting their tactical plans but also with re-examining their approaches completely. These incumbents have woken up to the fact that such experiments are necessary. If they fail to examine potential disruptions and continue to stick with the incremental evolution of the status quo, they stand to run into formidable challenges from upstart entrepreneurs and digital giants. Over the last three decades, GE's success has resulted from being a conglomerate that accessed and distributed its financial capital across a widely divergent portfolio of business to deliver superior returns to its shareholders. GE, under the leadership of Jack Welch, was in a class by itself. It succeeded wildly while many other conglomerate organizations were breaking apart. Under its successors, GE is reinventing itself as a digital industrial company; its experiments are at the intersection of physics and data, where its historical competency with materials science is now coupled with data sciences. GE was developing industrial IOT products connected to smart machines that collect data and use powerful software to analyze, optimize, and customize the product's usage for individual customers.

In general, tech entrepreneurs have specific vertical focus targeted at a particular industry and problem, sometimes experimenting to help industry incumbents and sometimes challenging the traditional ways. Incumbents are looking to complement and extend their traditional strengths in an industry or discard an outmoded business model and reinvent themselves for the future.

DIGITAL GIANTS: EXPANDING THEIR INFLUENCE ACROSS INDUSTRIES.
Digital giants approach experimentation at the edge differently. They have scale and scope ambitions to differentiate themselves from their direct competitors, other digital giants. So their experiments typically involve how they can achieve increased scale by newer means or how they can expand their scope into new domains.

Health care is increasingly interesting as a field in which digital giants are innovating and broadly expanding their scope. Almost every digital giant has at least one health related project in an experimental stage. Google might have abandoned its previous incarnation of health, a place to store consumer health data. Under its new governance structure, Alphabet has dedicated an entire unit, Verily Life Sciences, to exploring the question of how to use technology to create a true picture of human health. It has established multidisciplinary teams comprised of chemists, engineers, doctors, and behavioral scientists to work together to understand health and ways to prevent disease. Alphabet has joined with a Johnson & Johnson unit to create Verb Surgical, a company designed to create the future of surgery at the nexus of machine learning, robotic surgery, instrumentation, advanced visualization, and data analytics. It is a high-tech company to redefine surgery as it's practiced today. And Alphabet, with its focus on health, life sciences, and longevity will be a force to be reckoned with in the coming decade. Apple has taken a different approach. Its HealthKit is smartphone software that allows developers to create a wide range of health and wellness apps with sensors built into the iPhone and Apple Watch. Its ResearchKit develops apps that allow researchers to collect robust medical-grade data for research purposes, and its CareKit develops apps so that individuals can take a more proactive role in their fitness and well-being. Indeed, Apple's health-related software frameworks have sparked extended experiments in health care labs and medical research institutions. Time will tell how far Apple's software kits influence the future of healthy living.

RESPONSE: CURIOSITY AND COMMITMENT. Looking at experiments at the edge of your industry and others from the perspective of digital giants, tech entrepreneurs, and your other industry incumbents gives you a better sense, not only of the wide-ranging ways in which companies are testing their business models and the market, but also of strategies that motivate them. The leaders of all these companies share one thing: a curiosity to explore the frontiers of digitalization, where none of the previous ideas of product definitions, industry boundaries, or value propositions matter. Leaders of successful digital companies see fascinating possibilities and carry out experiments to develop promising pathways to create their digital future. These are not technical experiments but business ones. They are not single, isolated initiatives, but sequenced and coordinated as a portfolio of interdependent ones.

Go beyond just passively observing digital trends and undertake your own tactical experiments to get your feet wet and test the water. Be broad, observe and commit. This is where digitalization becomes real and intense.

COLLISION AT THE CORE

Sooner or later, incumbents in every industry are going to collide with tech entrepreneurs and digital giants, who have a different idea about how to design and deliver products and services, and how they can earn revenue and profit. Experiments at the leading edge can move very quickly, and understanding how this happens can help companies be prepared.

SIGNS THAT A COLLISION IS TAKING PLACE. Have you started noticing companies from outside your industry beginning to challenge your business model? Have you seen innovation in value delivered to customers that take advantage of digital technologies? Have you seen experiments from yesterday evolve and become well-thought-out business ideas that challenge the traditional ones? If you find yourself in this phase, or if you believe that it is imminent, digitalization should be on your management agenda as a fundamental driver of growth and profitability for your company. Businesses typically start to take notice of the need for digital transformation when a collision occurs because it's hard to miss the competitive pressure that threatens your very existence. You will feel this pressure first in one of two areas, and then both.

There are two kinds of collisions: strategy and organization. A strategy collision means your traditional strategic logic, such as your business model and competitive strategies, collides against newer reasons that rely more heavily on digital functionality. The newer models could come from other industry incumbents who have accelerated their digital transformation to form newer digital-born entrepreneurs or digital giants. An organization collision means your organizational model—including traditional structure, processes, and systems born in the industrial age—collide against alternative ways of organizing born in the digital age. Both of these types of collisions are increasingly interdependent and occur together. Moreover, those companies born digital do not see any compelling reason to separate strategy and organizational models into functionally defined boundaries or hierarchies, or even within and across corporate boundaries. To them, the information officer is the very essence of why they exist and how they plan to win. There is no separate chief digital officer or separate digital innovation unit. Any organizational structure today is irrelevant because no competition or innovation is going to respect those boundaries. Everything now is going to have to be much more compressed in term of both cycle times and response times. In other words, even companies that are large and successful digital giants, such as Microsoft, are facing this collision at the core of their business models. Microsoft's past strategies and structure as a packaged software company will not withstand competitive threats from other digital giants and tech entrepreneurs whose business logic is defined by mobility and cloud computing. Microsoft must reinvent itself when faced with this collision.

HOW CAN YOU EMBRACE DIGITAL TECHNOLOGY? It is not that incumbents are unable to adapt. It is that, ultimately, transformation depends on the level of commitment that your business makes to digitalization. It means investing in new competencies in hardware design, software, and applications, and making new acquisitions and new sets of alliances and partnerships. Equally important is recognizing the new organizational capabilities you need to compete against companies born in the digital era.

Let's examine the automotive industry. The question is if a "car", or a "computer on wheels" is connected to the cloud? In the changing world of transportation, what is an automobile anyway? Is it the industrial combustion engine that we first called a horseless carriage in the late 19th century? Or is it a network of computers with tens of millions of software codes connected to the cloud? Which one is a more appropriate way to think about this vehicle in the digital era? Framing a car as a physical product with services wrapped around it limits our thinking to the way the automobile industry functioned as linear value chains until the end of the twentieth century. By visualizing it as an intersection of software, entertainment, telecommunications, and cloud connectivity, we see the beginning of a mobility and transportation ecosystem where the auto sector interconnects with other industries to deliver broader value propositions around mobility and transportation. As much as a smartphone can be far more than just a tool for communication, a smart car can be more than just a means of transportation. Mercedes-Benz saw how different industries could morph into mobility and transportation ecosystem, and it already had a partnership with Google to bring Google Maps and Directions, Street Views, and related search programs into the dashboards of Mercedes cars. These experiments were intended to provide complementary services to Mercedes drivers with telematics. Google and Mercedes collaborated to co-create new value for customers.

As automobiles become computers on wheels, especially as more automobiles are manufactured with hybrid and electric drivetrains, Google has sought to have greater influence. Google has extended the scope of its Android software beyond mobile phones to cars, and beyond dashboard maps. This mobile app offers drivers control over GPS mapping, music, messaging, phone calls, and web searching, using touchscreen, button controls, or voice commands. More importantly, Google has successfully demonstrated its self-driving car project. Google and Fiat have entered into a preliminary agreement to collaborate on next-generation automobiles.

On the other hand, traditional automakers also joined the disruption of the transportation industry. If digital competencies become the differentiators and automakers do not excel in them, they could be pushed to the commodity end of the ecosystem. We may not yet know the dominant operating systems that will drive the automobiles of 2025, but we do know that digital giants are capable of strongly influencing this industry as it races towards digitally-defined, multimodal transportation.

The automotive industry illustrates the kinds of questions that will apply to nearly every industry as products get digitized and apps become central to connect and interoperate products on networks. The key collisions involve the processes and practices of how industrial age companies function. As an industry incumbent, the challenges are to rethink traditional product-centric decisions against the broader landscape of industry solutions that may become more central in the coming decade. The challenge is to learn to work cooperatively and competitively with other incumbents, and with the tech entrepreneurs and digital giants.

RESPONSE: COEXISTING AND THEN MORPHING. When digital business models first appear on the scene, they are fuzzy, ill-informed, and inelegant, most incumbents do not see their differentiating value or perceive their competitive threats. Indeed, research over two decades has shown that incumbents do not recognize disruption and fail to develop effective response. However, coexisting and morphing are two levels in figuring out effective timely responses.

Coexisting means you are setting up digital and traditional business models to coexist within the same structure. In Netflix, the old DVD-by-mail model and the new video-streaming model coexisting is the best way to phase out their historical business model. There will be the usual challenges: insufficient resources allocated to the innovation units to develop and grow, inadequate capabilities to efficiently compete against newer startups and digital powerhouses, and diffused management attention across traditional and new units resulting in neither succeeding. While you are incubating a digital business within the traditional structure, examine three avenues, including sharpening your value-to-cost advantage, examining your alliance advantage, and evaluating your acquisition advantage. You want to clearly and sharply delineate the distinctive value propositions of your existing model to preserve your revenue and profits. At the same time, recognize that the digital alternatives should not be underrated or minimized. Maintain your traditional business model at the core, but experiment with innovative digital alternatives as important satellites. In this stage traditional industrial business models coexist alongside new digital business models within your company, and certainly within your industry.

In the early stages of digital transformation, the traditional core gets more attention than the digital counterpart, but over time, as you decide to commit more resources to your new digital model, the core of your business will morph form traditional to digital. To make the best use of the window of transition while you are morphing your core, there are three principles to follow: To divest from traditional business to focus on your new digital core; to absorb digital acquisitions into your core, and to instill digital business thinking at your core. Over time, the hybrid model of traditional industry and newer digital models coexisting will prove limiting with competing pulls and misalignment with market demands. Successful organization will effectively morph into a fully digital business model. Once you have shed inefficient ways of working, you start to resemble the tech entrepreneurs, but with scale. You start to see the power and capabilities of digital giants. And you begin to see a future in which your domain knowledge and detailed subject in your industry, combined with digital thinking, will give you as much right and competence as the other players to be the leader in the next phase.

REINVENTION AT THE ROOT

"What business are you in?" The answer to this question in the digital era is that new business models emerge at the intersection of industries. An organization must learn to think of itself not as producing goods or services, but as doing the things that will make people want to do business with it. To accomplish this, you need to focus not just on the design and delivery of products and services, but on solving problems and shaping solutions.

Instead of "what business are you in?", the two fundamental and interrelated questions to ask for reinvention are: First, "What problems are we trying to solve for consumers in the world?" and second, "How are we uniquely solving them by taking advantage of digital technologies?" These two questions compel you to look at the way digital technologies have altered the core elements of the traditional business model and shifted the activities from one company to another in the broader ecosystem. They also compel you to examine how new interactions could be effectively monetized.

WHAT PROBLEM COULD YOUR BUSINESS DISTINCTIVELY SOLVE? The key point is that such problems have to be solved at scale, across industries, and at speed. These are not high-touch, high-fee, personalized, bespoke solutions for a few wealthy segments, these are high-value, highly innovative solutions for large segments of society. So instead of defining the business you want to be in, focus on the thorny problems that you want to solve.

Industrial Internet. GE is seeing bold new possibilities for problem solving. Long known as a home appliance company, it is reinventing itself by combining its traditional competencies in materials science and physics with new capabilities in data, software, and analytics to frame and effectively solve core problems in several sectors. For example, by using machines with sensors and software connected to the network, GE can observe, use predictive analysis, and provide proactive solutions to enhance efficiency in factories. With real-time visibility, the end-to-end manufacturing processes can be tightly controlled, and maintenance can be accurately predicted and carried out when it's needed and when it's convenient rather than on a predetermined schedule. So, GE solved its customers' inefficiency problems in ways that were impressive without digital technologies.

Other companies in the industry such as Bosch, Siemens, and Samsung are now also using data and analytics to find and eliminate their own pockets of inefficiency. The result, which is called Industry 4.0, is an increase in productivity and a decrease in cost and time delays across different companies in the value chain. Such reinvention at the root of manufacturing and supply chains could save many billions of dollars over the next decade. The invisible domains of digitalization, manufacturing, supply chains, and distribution, become as important as the visible domains of digitalization, the customer-facing interactions with apps, ads, and algorithms.

Business solution. IBM's reinvention marks a significant shift from its historical role as a systems integrator, meaning that it tied together the disparate pieces of systems, such as computing hardware, software, and services, better than other competing integrators or than its customers could do by themselves. The reinvention positions IBM at the intersection of technology and deep domain knowledge to address questions such as, how could cities function more effectively given massive urbanization, how could quality health care be delivered affordably in critical areas such as cancer, and how could big data and analytics seamlessly enhance knowledge? IBM is positioning itself as a solution integrator, meaning it brings relevant knowledge based on its extensive experience across industries to uniquely solve each client's unique business problems. It has divested the company from its low-value hardware products to focus on high-value areas, accelerated its investments in cloud and artificial intelligence technology, and created a separate unit to commercialize the cognitive computing frontier. It has also made significant acquisitions, such as The Weather Company to shore up its domain knowledge in specific verticals and created global alliances with companies such as Apple, Microsoft, and Cisco to expand its scope and scale of expertise.

WHAT BUSINESS MODEL IS RIGHT FOR REINVENTION?

The digital transformation involves a shift in thinking that leads you from a product or service-focused business model to a problem-solving one. To unpack this shift, it can be helpful to look at four archetypical business models, starting with two that will already be familiar to you.

Product. These are the tangibles offered for sale that we know from the Industrial Age. They are computers, refrigerators, washing machines, and light bulbs. In the digital era, companies still make the same products, but they are fitted with sensors and software to capture data and link them to other products and services. Products get smarter as they are digitalized; for example, cars started with telematics and have now become computers on wheels connected to the cloud.

Service. These are intangible products, or the actions that are performed to fill a need or satisfy a demand. The services such as hotels, banking, entertainment, and education from the industrial era, but digital-era services are supported, shaped, and delivered by digital technologies. Services get smarter as the data about them get richer. Whereas traditionally automakers had a lot of data on cars up until the point of purchase (the ownership record), which allowed them to produce the best cars compared with their incumbent competitors, newer service companies such as Uber, Lyft and Waze define their mandate more broadly. They collect reliable, detailed data on how cars are used (driving record) or related pain points (parking, traffic) or auxiliary services (insurance, maintenance) so that they can deliver the best transportation service from A to B at specific times. This more detailed outside-in information about customer expectations and experience is much richer than the old inside-out information derived from market research data.

Platform. These are the computer operating systems, video game consoles, smartphones, search engines, and more that have risen in the digital world. Platforms connect many different types of companies transacting with many different types of customers. As platforms grow by digitally connecting more individual companies, the value to consumers of their products and services is increased because they work together in a way that none could achieve on its own. In the case of the automotive industry, these are Apply CarPlay and Android Auto for now. They seamlessly extend services from smartphones to cars, but their scope could expand further in the future.

Solution. These customizable products, combinations of products, services, or mix of products and services, have become more prevalent in the digital world, in which companies are better positioned to solve specific business problems through data and analytics. For example, we are on the cusp of seeing digital technologies, such as the conversation bots that observe and understand you and then recommend precisely what you need, delivering a personalized solution at an affordable price. In the automotive sector, GM's Maven is beginning to offer personalized car-sharing services. Daimler's Moovel is reinventing the future of urban mobility, using technology to connect modes of transportation so that individuals can travel in different ways, including walking, biking, riding public transit, and driving, based on what's convenient and easy to use.

RESPONSE: PROBLEM FRAMING AND PROBLEM SOLVING, TIED TOGETHER.

What is common in this reinvention phase is a shift in business logic. Incumbents are reimagining their business models beyond products and services, tech entrepreneurs are introducing new business innovations rooted in powerful technologies, and digital giants are extending their platforms and offering relevant integrated solutions. As you determine what differentiates your company, ask: What is your relevance when business logic shifts from delivering products and services to solving problem and shaping solutions?

Frame the problems from the outside in. It is easy and comforting to simply define the problem in ways that match what you have offered, but think like a consumer instead. Step into the shoes of the customer and understand their pain point, their needs and frustrations in their daily tasks. Be sure when you frame your business problems that you are looking deeper than the domain of products and services to the different pain points for customers. And those pain points may not fit neatly into traditional industry definitions. The biggest challenge facing incumbent companies is one of reorienting from the familiar "product push" to the unfamiliar "customer pull." In other words, embed digital technology in your interactions with customers to better understand their needs. If you can quantify and analyze how your customers arrive at their decisions, you can better frame the right question, tailor the relevance of your solution, and succeed over those who still tell customers what they want.

Problem solving transcends industry and disciplinary boundaries. Having framed a problem, you need to determine a rational, realistic, and systematic way to go about solving it by bringing the power of digital technology to bear in way that could not be done before. It's well known that innovations are complex, but it is also known that creative solutions emerge when you can transcend the traditional disciplines and industry boundaries. Companies have used crowdsourcing to tap into the wisdom of groups to solve problems. For example, P&G has relied on open innovation in its "Connect + Develop" program to tap into intellectual capital and expertise that exists beyond its corporate boundaries. Artificial intelligence, machine learning, and cognitive computing—all overlapping ideas—look for patterns that lie across industries and disciplinary boundaries, which is an exciting frontier for solutions delivery with predictive analytics at the core. Meanwhile, individual companies do not solve big thorny problems on their own. You may be able to solve part of the problem by tapping the collective expertise in your own firm, but only by working with other incumbents, entrepreneurs, and digital giants are you going to solve the larger problem. You must focus not only on *me* but also on *we*. So, cultivate a network of relationships and delineate the distinct roles each player has within the ecosystem. Be open to new ways of thinking about problems as you interact in the ecosystems and find creative ways to solve problems.

KEY ACTIONS ACROSS YOUR THREE PHASES

By now, you have seen the individual meaning of the digital transformation agenda, and also seen how the nine charts are interrelated. You can see the three players and three phases intersect in complex and dynamic ways, and you can understand better how they feed the scale and scope at speed in the digital economy ecosystem. You may have sensed from many examples that the winners have been using some powerful strategic moves to navigate this fluid arena and succeed across the three phases. You see how the key actions flow linearly: from observe to invest in phase one, coexist to morph during phase two, and problem framing and problem solving in phase three. Yet you also understand that those actions have rapid feedback across the phases. These actions across the three phases are intertwined and give rise to effective and efficient strategic moves for your digital transformation. Three such winning moves are our focus in our subsequent discussions.

PART TWO

BEING DIGITAL

CHAPTER 4
MOVEMENT ONE: FROM PLATFORM
TO ECOSYSTEM

What business are you really in? The fundamental question has become even more important for the business of being digital, as companies are moving from products to platforms, and even to ecosystems.

EXPANDING BUSINESS SCOPE AND SCALE IN A DIGITAL ECOSYSTEM

Online retailers. When Amazon first launched its website in 1995, the goal was to use the Internet to sell books at low prices. Amazon created a virtual store with lower fixed costs and a larger inventory than most retail bookstores. The concept quickly became popular, and Amazon realized that consumers might also appreciate this concept when they shop for other types of goods. So, Amazon began adding dozens of categories to their online assortment, including music, DVDs, electronics, toys, software, home goods, and more, which was a threat to retail giants such as Walmart and BestBuy. Five years later Amazon opened its site to third-party sellers, who could post their products on Amazon for a modest service fee. This was a win-win deal because third-party sellers increased Amazon's assortment without extra inventory, and sellers got access to the increasing pool of consumers who enjoyed shopping on Amazon. Adding third-party sellers also transformed Amazon from an online retailer to an online platform, which required Amazon to develop new capabilities of acquiring, training, and managing sellers on its sites, without losing control or damaging customer's experience. And its competition set expanded to include eBay.

Streaming content. Recognizing the changes of consumer behavior—downloading digital music instead of buying CDs in a store—Amazon launched its video-on-demand service, following its customers and shifting from selling CDs and DVDs to offering streaming services that required it to develop new capabilities and pit it against a new set of competitors such as Apple and Netflix. In 2011, in partnership with Warner Bros, Amazon launched Amazon Studios to produce original motion picture content. Suddenly it was competing against Hollywood studios. Why did it make sense for Amazon, which started as an online retailer, to move in this direction? Because video content helps Amazon convert viewers into shoppers, and the original content of Amazon Studios also encourages Prime members to renew their subscription. Launched in 2005, Prime offers free two-day shipping—and even one-day or two-day—for a subscription fee. By 2017, Amazon had almost 75 million Prime members worldwide. Not only does the subscription fee generate almost $7.5 billion in annual revenue, but Prime members also spend almost twice the amount of money that other Amazon customers do. In addition to creating loyalty among Prime members, original content is also a means of attracting new customers, and Amazon invested $1.3 billion in original content as a key driver for acquiring new customers in other parts of Amazon's business.

The Hardware business. In 2007, Amazon released the Kindle to enter the hardware business. The Kindle was designed to sell eBooks as consumers shifted from physical products to digital goods. The Kindle, like a razor, is selling at a low price in order to make money on eBooks, which would be akin to the blades. More recently, Amazon launched additional devices: Dash buttons, which let users order products from over a hundred brands when users' supplies get low, and Echo, a voice-activated virtual assistant, which can be used to stream music, get information, and order products from Amazon in an even more convenient fashion. Echo was launched in 2014, and within two years Amazon had sold almost 11 million Echo devices and had built over 12,000 apps or skills for this device.

Advertising networks. Amazon also started its own advertising network, which put the company squarely in competition with Google. Amazon's large customer base, and more specifically the company's knowledge of consumers purchasing and browsing habits, provides Amazon with a rich source of data to target its customers with relevant ads. This shift allowed Amazon to generate almost $3.5 billion of ad revenue in 2017. But an even bigger goal for Amazon is to replace Google as a search engine for products, so that customers start their product search on Amazon rather than on Google. This would not only reduce Amazon ad spends on Google, but would also give Amazon tremendous market power.

Web services. Understanding the digital business ecosystem and whether to participate or position yourself to orchestrate within it is an important winning move. It allows you to know where to apply your energy and how to make the most of your resources across relevant ecosystems; it also allows you to identify your competitors and your potential allies—and when—both inside and outside of your existing ecosystem. When Amazon entered the cloud-computing market with the launch of Amazon Web Services, suddenly a completely new set of companies–for instance, IBM—became Amazon's competitors. What is an online retailer doing in cloud computing? AWS helps Amazon scale its technology for future growth. It allows Amazon to learn from other e-commerce players who use its platform. And it enables Amazon to leverage and monetize its excess web capacity. AWS is a way for Amazon to build its technology capability to become one of the largest online players and monetize that capability at the same time. In the fourth quarter of 2017, AWS generated over $5 billion in revenue, representing annual revenue of more than $17 billion and 43 percent year-over-year growth.

From platform to ecosystem. As an online retailer, Amazon competes with Barnes & Noble, Best Buy and Walmart. As an online platform, Amazon competes with eBay. In cloud computing, it battles for market share with IBM, Google and Microsoft. In streaming services, it has Netflix and Hulu as formidable competitors. Amazon Studios puts the company up against Disney and NBCUniversal. Its entry into digital devices puts it in the crosshairs of Apple and Samsung. Its ad network makes it Google's rival. Most companies define their business by either their products or their competitors. You may consider yourself in the banking business or the automobile industry. But it is hard to define Amazon with this traditional viewpoint. Amazon expanded its scope all around the business ecosystem.

It should be clear that competition is no longer defined by traditional products or industry boundaries. The rapid development of technology is making data and software integral to almost every business, which is blurring industry boundaries faster than ever before. The smart, connected devices, or the Internet of Things, shift the basis of competition from the functionality of a single product to the performance of a broad system. In today's connected world, sustainable competitive advantage comes from offering a system of connected and complementary products, and from creating a platform with strong network effects that increase consumers' switching costs.

Meanwhile, a broader business scope requires building new capabilities. There are two ways for business expansion: Take inventory of what you are good at and extend out from your skill, or determine what your customers need and work backward. Traditional giants expand into adjacent business where they can leverage their existing capabilities. However, a customer-centric view requires a firm to follow shifts in customer needs and to develop new capabilities to meet those needs.

SELECT A BUSINESS MODEL AND STRATEGY IN A DIGITAL ECOSYSTEM

A business model defines the way a firm creates, delivers, and captures value. Technological innovations lead to changes in consumer behavior and to emergence of new competitors, requiring a company to transform its business model to survive and thrive in the digital era. There are four archetypical business models including "Product", "Product + Service", "Product + Service + Platform", and "Product + Service + Platform + Solution". Each model offers a distinct expertise and delivers different values in the ecosystem.

To illustrate these four types of business models, suppose that a light has burned out in my house. What business models are necessary to solve my problem?

Products. In most case a standard product (a light bulb) may suit the purpose. Over the years, light bulbs have been designed and manufactured in standard formats and shapes and distributed through traditional outlets. So, I go to my local hardware store, buy a light bulb and install it myself.

Product + Service. A simple but personalized service might be replenishing my light bulbs through a subscription service. Here the manufacturer, or my local store, may have more information about the type of light bulb I use and how often I replace them and either alert me to the fact that it will soon be time to replace one or know exactly what I need when I call to buy one.

Product + Service + Platform. Let's say I decided to connect the light bulbs in my home to my alarm system and my sound system via an app on my smartphone. I want my lights to turn on as soon as I disarm my alarm and open the front door, and I want my preferred music to come on at the same time. My light bulb is no longer just a standard standalone product that operates on a stand switch. It is now connected to Wi-Fi and the Bluetooth network, and to my alarm and my music player, as part of the IOT. So now I am dealing with a product, a platform, and a service. I probably could go to my local hardware store to buy the light bulb, but chances are I now order it online or just press the "dash button" to reorder directly from Amazon. I reorder the light bulb using my smartphone; wait for it to be delivered, and install it myself.

Product + Service + Platform + Solution. I still need the product (light bulb), the platform (Apple Home Kit), and the service (retailing). And I could do the research, order the light bulbs, and install them myself. Or I could specify my requirements to a solutions architect (for example, a smart home technology designer), who will source the best individual product (the light bulbs) with the appropriate service (for example, design and maintenance) and connections to my existing home-networking platform. Here we have got four or more participants, with four different but complementary business models, delivering value to me, the customer. In other words, these four business models represent four ways of solving the problems that I, and other customers, have. Moreover, these four models are interdependent. In some cases, companies with different business models work together to create more value for the customer. In other cases, one model provides a greater share of the value.

The platform model seeks to maximize the number of players and offerings that are linked to it by broadly interconnecting different products and services, and it seeks to increase the scale of adoption by consumers. The solution model seeks to pull together the most relevant set of players and offerings and integrate these specific pieces to solve the unique needs of particular customers and generate revenue and profits by doing so.

The strategy of Razor and Blade in a set. For a long time, the music industry enjoyed significant growth by selling music in physical formats: Vinyl, cassette tapes and CDs. Remember the days when you used to buy a CD with twelve to fifteen songs, even though you really wanted to listen only to one or two? Why did the music industry force you to buy twelve to fifteen songs? The cost of producing, shipping, and selling a CD did not change significantly whether a CD had one song or twelve songs, so it made sense for the industry to bundle multiple songs and charge a higher price to cover these costs. However, digital technology changed all this. The cost of reproducing and distributing music dropped significantly in the digital world and allowed iTunes to sell individual songs

While the unbundling of music was good for consumers, it wreaked havoc on the music industry, which saw a significant drop in its revenues. The sale of digital singles and even revenue from streaming services did not come close to equaling that which came from selling bundled music on CDs. Music piracy exacerbated the problems for the industry. As a result of these dramatic shifts, those in the business of making music such as songwriters, musicians, and sound engineers, also saw their incomes take a nosedive. The industry had arrived at a kind of paradox: while the popularity of music among consumers reached record highs, the music studios and artists suffered a dramatic drop in their income. Music studios and artists had traditionally used concerts to generate awareness and excitement among fans to sell and make money on music albums. In other words, concerts were the razors to sell music albums—the blades.

Companies have used the razor-blade strategy for a long time: sell razors cheap to make money on the blades. As the income from the selling of recorded music (the blades) declined, studios and artists converted the razor, such as concerts, into blades, or the money-making part of their business. Suddenly free and even pirated music becomes the cheap razor to drive fans to the expensive concerts. Artists have benefited from this shift, since the percentage they generally receive on concert revenues exceeds the royalties they receive from the sale of music albums. Artists have also been able to exploit the popularity of their music by entering direct partnerships with brands. Companies such as MasterCard offer their customers unique experiences by inviting music artists to perform at special events that have become a lucrative source of income for the artists. Global revenue from live music performances is expected to be more than $31.5 billion by 2023.

Most companies have multiple sources of revenue, often from complementary products. Complementaries provide a new source of competitive advantage in the connected digital world by increasing consumers' switching costs. Even if you think your iPhone is similar in its product features to a smartphone from Samsung, you will find it hard to switch to Samsung if you use Apple complementary products and services, such as iTunes, Face Time, and Apple Pay. And complementary offers provide another key advantage: as technology affects a company's profitability. It can transform the company business model by shifting revenue sources from one product or service to another, effectively converting a razor into a blade.

SEEK TO ORCHESTRATE OR PARTICIPATE ACROSS ECOSYSTEMS

In today's digital economy, ecosystems are an essential part of doing business, every company must now ask itself, are we seeking to orchestrate the ecosystem or simply to participate?

In Industrial-Age industries, the leaders are the companies that are best at vertical integration. They own the key assets from end to end, whether it is a product or a service. In the 1980s, IBM was a classic example because its mainframe computer systems were made up of its own proprietary hardware, software, and services, which came only as a total package and did not work with any other company's computer system. But in the digital age, the leaders are the companies with the best virtual integration; they control the market by assembling and managing the best network of product and service providers. Microsoft was a great example when it introduced the Windows operating system. Instead of keeping it a closed system, like IBM and Apple, they made it available to other company's products. In other words, whereas IBM provided a complete product, Microsoft provided a platform, a proprietary wedge of software between the hardware and the different applications, to which other companies could add their products and generate their own profits. Microsoft, along with its network of hardware and software providers, developers, and retailers, providing products and services for its platform, constituted an ecosystem.

The ecosystem does have a hierarchy, with definitive leaders and followers. The leaders in the ecosystem are orchestrators and the followers are participants. Both are needed for an ecosystem to thrive, grow, and win. Orchestration is about pulling together companies with different business models and strengths in different industries and connecting them across traditional industry boundaries. This is where digital giants have intrinsic strength and superiority. Orchestration is about making links between manufacturers, service providers, platform providers, and solution architects to create a system of network effects. Participation is about knowing your core strength and allowing others to link to you, creating value greater than would otherwise be possible alone.

Orchestrate to lead the ecosystem. IBM earned its revenue and profits as a system integrator of end-to-end proprietary architecture in mainframes during the 1970s and 1980s. In contrast, by the time personal computers were widely used in homes and offices, Microsoft earned its revenue and profits as the orchestrator of the Windows-Intel personal computer ecosystem in the 1990s. Microsoft, despite controlling only a small part of the end-to-end assets in computing, shifted the focus of value from IBM to an ecosystem of complementary players that delivered a wide range of products, services, and solutions on a Microsoft platform. It designed the Windows platform with Intel. It partnered with Dell, which had perfected how to build to-order PCs with superior supply-chain capability, and even without its own physical stores had emerged as a leading reseller of the Wintel platform. It worked with HP and Epson, and Kodak, whose scanners and printers provided PC users with enhanced value. In essence, through its Windows OS in the software layer and Office Suite in the applications layer of computers, Microsoft managed all of the relationships in the design, manufacture, sales, and service of PCs, from chips (Intel) to services and solutions (Accenture).

The industry's main business logic shifted from selling computers as hardware (a product) wrapped with services to selling computing solutions through platforms (with complementary products and services) for customized applications and use. Clearly, individual products such as scanners, printers, software applications, and services like configuration, system integration, and consulting were still necessary, but Microsoft created a new pocket of value by linking them as part of broader platforms and solutions. Customers benefited, and all of the players on the platform earned revenue and profits they could not have earned on their own. Designing its personal computer operating system in ways to attract complementary players to create this system of network effects enabled Microsoft to earn both a significant share of the value as the platform's architect, and the right to orchestrate the personal computer ecosystem.

Why Microsoft failed to capitalize on its initial advantage over Windows and the personal computer to expand its ecosystem to include mobile, which Apple entered only in 2007 and has dominated over the last decade, or search, Google entered in 1998 and has dominated ever since, or social, which Facebook entered in 2006 and has dominated ever since, is a story for another day, but it underscores the fact that ecosystems are dynamic. As we saw earlier, future success is never guaranteed by past success. Although Microsoft orchestrated the PC ecosystem and continues to do so even today, its role in mobile, search, social, and other emerging areas is more like a participant.

Participate to support the ecosystem. Every ecosystem has several participants with different but interconnected and interdependent roles. An ecosystem is powerful and vibrant when supported by many different types of participants whose offerings complement each other. In other words, as a participant, you must understand that the value you deliver today may not be the same value you could deliver in the future. Choose your business model wisely and evolve as conditions change.

Now let us look at retail payment ecosystems, which are transitioning from magnetic and chip-and-PIN cards towards software apps inside smartphones. The transformation cannot happen overnight, because of the legacy infrastructure involving terminals and card readers, but digital payments are clearly on their way. Here again Apple and Google emerge as potential orchestrators, with Apple Pay and Android Pay. There are also the traditional players, such as issuing banks (Chase, Bank of America, Barclays), card networks (American Express, Visa, MasterCard), acquirers and processors (Chase, Citi), payment gateways (PayPal), and a wide array of merchant service providers. How could the digital payments systems evolve over the next decade and who could orchestrate them? Should an incumbent company such as Walmart, participate in the retail payment ecosystems proposed by Google and Apple? Should Walmart participate in both ecosystems? Does Walmart have unique skills and capabilities to explore a preferential relationship in one ecosystem over another? Are there opportunities to further leverage Walmart's own digital capabilities to orchestrate mini ecosystems that could link to both Apple and Android?

When Apple introduced Apple Pay in 2012, Walmart would have answered no to all four questions, mainly because it wanted to orchestrate digital payments, at least around digital commerce, online and in stores. Long before Apple and Google announced their intent to move into retail payments, Walmart had pulled together a consortium of merchants in an ecosystem with a vision to create Current C. It was positioning to orchestrate its own payment ecosystem and had strong backing from retailers such as 7-Eleven, Target, Best Buy, and Lowe's. So, Walmart's initial response to Apple's payment ecosystem was to reject it in favor of continuing to show commitment to the Merchant Consumer Exchange and its projected Current C payment system. As a would-be orchestrator, Walmart persuaded the member community to reject Apply Pay; it wanted to prevent its participants from defecting to this competitor ecosystem. The participants were divided about what to do. Some merchants decided to shut out Apply Pay in their store but others wondered whether to continue to participate in this initiative exclusively or to renegotiate the contract to explore working with Apple Pay. With more options to participate in than before, some merchants wanted to choose the system with the broadest scale, customer appeal, ease of use, and reliability. They were no longer interested in Current C, which appeared to be an inferior system after the introduction of Apple Pay. In late 2015, Walmart continued to support the collaborative initiative, but concerned that Current C might not gain the necessary support of other retailers, it introduced its own proprietary Walmart Pay as a competitor to Apple Pay. Walmart's standalone currency may gain traction within its stores and with its customers, who could be lured with attractive offers, but for the company to credibly orchestrate its own retail payment ecosystem, it needs more participants. But so far, it's only leading its own closed payment ecosystem with very little participation from the other retailers.

In the digital business world, every company is embedded in multiple ecosystems that can be understood using the four archetypical business models. It is important to recognize you could well be a member of more than one ecosystem, which means that it is possible for you to be the orchestrator in one and a participant in others. The principles that give you a working guideline to decide where you could credibly step up to orchestrate and when to provide support to someone else to be the orchestrator are: Define your relevant set of ecosystems, decide on your role in each ecosystem, and examine the dynamics of these ecosystems.

Understanding digital business ecosystems and whether to participate or position yourself to orchestrate is an important winning move because it allows you to know where to apply your energy and how to make the most of your resources across relevant ecosystems.

CHAPTER 5
MOVEMENT TWO: FROM
COMPETITION TO "CO-OPETITION"

THE RISE OF OPEN INNOVATION

Companies spend billions of dollars on R&D with the hope of creating innovative products that will give them a sustainable competitive advantage. Most business managers assume this producer or company led model to be the dominant model of innovation. However, research found that most users engage in product development out of personal need to adapt and modify existing products for their own specific use. As product lifecycles become shorter and R&D costs continue to increase, internal innovation alone is no longer enough to support companies' growth expectations. To remain competitive and hit topline growth goals, and to extend the reach of their innovation pipeline and leverage their limited resources, many companies have found open innovation to be a necessity.

Technology has also played a significant role by making design, development, and collaboration tools affordable and accessible, which in turn has made innovation possible for small businesses, communities, and individuals. Firm boundaries become less rigid as transaction costs go down, making it easier for outside players to provide input to a company without being employees. Technology has also enabled like-minded people to gather in large virtual communities where they share information and brainstorm ideas.

Internal company teams often view a problem through a single lens and attempt to find a solution using a handful of approaches. In contrast, open innovation casts a wide net and attracts large numbers of participants with various areas of expertise who employ a range of methods and perspectives to solve a problem. In 2006 Netflix created a competition for a one-million-dollar prize to anyone who could improve its algorithm for movie recommendations. There was great insight among some of the teams. People participating in open innovations are usually part of a large community, they are highly competitive, and often share their winning entries with others, which provides a strong learning platform for participants to improve upon in the future. Meanwhile, user innovators are by definition close to their market, as they are, in essence, innovating for themselves. These users are often working to meet their own needs before firms even identify the opportunity. By default, this puts user innovators ahead of firms on the innovation timeline.

To successfully leverage the power of open innovation, firms should consider the following issues: Defining the problem, breaking it down into components, and integrating the solutions to these small problems, creating a clear metric for evaluation, designing the challenge, and managing the challenges. Open innovation also requires firms to give up control and broaden their lens. Often ideas come from a very different field, and it is tempting to reject them out of hand. In a crowdsourcing future, it will be the questions companies ask, not the solutions they know, that will determine their success.

THE OMNI-CHANNEL STRATEGY

As the world moves from bricks to clicks, companies are struggling to develop an effective Omni-channel strategy. By now everyone recognizes that the choice is not whether to have physical stores or digital channels, but how to manage both at the same time. How do you avoid channel conflict? How should you link digital and physical channels? Should physical stores be redesigned as digital channels evolve?

Channels are complements, not substitutes. Within the discussion of channel conflict is the assumption that different channels compete against one another as substitutes. To some extent these beliefs are true, and in the long run online transactions may indeed replace a large fraction of physical transactions. However, the challenge is how to manage the transition. How do you build the online channel without jeopardizing the large and often profitable business from the offline channel? The key to managing this transition is to think of different channels as complements, not as substitutes. Each channel is best suited for certain products, for a specific group of customers, or for a certain part of a consumer's decision journey.

With the current consumer behavior, the online channel was perhaps best suited for selling simple products. With its extensive reach and low cost of marketing, the online channel could become an effective and efficient customer acquisition tool. And once customers were acquired through the online channel, they could be handed over to existing dealers, who could cross sell more expensive and complex products to them. This approach appealed to dealers, since it reduced their burden of customer acquisition, leaving them to focus on selling complex, higher-margin products that required their expertise. As the company gains experience in building its online channel and as customers become more comfortable in navigating through a complex array of products on their channel, the company expects to see an increasing migration of customers and of revenues from its physical distribution system to its digital channel. Once again, it helps to think about the role each channel might play for a different product. The online channel can offer more variety and customize products for consumers. Such offerings would not cannibalize retail sales. Instead, they would provide different and complementary products to a subset of customers who desire distinct, customized choices. And even though it is cheaper to serve customers through digital channels, the company could charge a premium for these offers, further reducing cannibalization and potential conflict with its retailers. Over time the digital channel would grow as the company gains more experience and as customer behavior evolves.

Different channels may serve customers at different stages of their decision journey. For example, boutique skincare brand Kiehl's utilizes engaging window displays that serve as expensive billboards to intrigue passersby and entice new customers. Once a customer walks into a store, a trained salesperson can communicate the brand value proposition to them and identify the best products that might suit that particular customer. In effect, these stores are a great vehicle for customer acquisition. Department stores and other retail partners generate lots of foot traffic and also help in extending the reach of the brand. Given the brand's low awareness, the company EC site is used mostly by existing customers looking to find new products or to order supplies of their favorite product. In other words, the online channel serves the best for customer retention.

Physical and Digital fusion. Talk to any retailer and you will hear of the importance of embedding technology, such as beacons, in stores. Beacons would capture data about consumer traffic patterns, data that would enhance retailer's ability to serve their customers. Yet there is hardly a retailer who has been able to leverage such data effectively. Starting with technology is rarely productive. Instead, ask yourself what consumer problems you are trying to solve and how technology might enable you to solve those problems. In a retail environment there are at least four consumer pain points that warrant attention: Finding things, trying things, paying for things, and returning things. As one of the value propositions from Amazon, the goal of a company should be to remove friction for consumers, and technology should be used to do exactly that. In recent years Amazon shocked the retail industry by opening a handful of physical stores.

Why would a highly successful e-commerce player open physical stores when it had gained a unique competitive advantage by eliminating the fixed cost associated with brick-and-mortar stores? There are at least four reasons for Amazon to test this Omni-channel strategy. First is for new product categories. Retail stores continue to dominate several product categories, such as groceries, furniture and large appliances. For such categories, consumers still prefer to shop in person. Second is for Amazon devices. In recent years Amazon has become a major player in devices, with such offering as Echo, Dash Buttons, and the Kindle, most of these products are designed as complements that help Amazon sell things such as books and other merchandise. Amazon has sold these devices through retailers such as Best Buy, and also in their own store, to increase visibility. Third is Prime membership. Since online commerce in the USA accounts for only about 15% of all retail sales, the offline market represents a huge untapped area for Amazon. Physical stores can potentially become a customer acquisition channel for Amazon Prime membership and online commerce. The last but not least is reinventing retail. Testing physical stores also allows Amazon to reinvent the retail industry.

It has already done so with Amazon Go. Further, it is using customer data and customer reviews to select and display books for its bookstores. In the future, it is conceivable that Amazon will sell retail technology service to other retailers, just as it built AWS first for its own e-commerce business and later offered it to other companies.

COLLABORATE TO CREATE NEW CAPABILITIES

Most people think about business as a competitive environment. You compete for market share; you compete to earn more revenue than our rivals. But there are also times when you have to cooperate. You cooperate with the suppliers and distributors, you cooperate with your customers, you cooperate with lawmakers and regulators. Digital business, with its emphasis on networks and working in ecosystems, blurs that distinction. You have to compete and cooperate at the same time. "What business are you really in? "The fundamental question has become even more important in the past five decades, as companies are moving from products to platforms, and even to ecosystems. The shape and structure of digital business ecosystems will change even faster; it will dramatically change the nature of your relationships within this ecosystem. Knowing when to transact, when to lead, when to follow, when to co-create is the key to your second winning move.

"Co-opetition" is a revolutionary mindset with competition and cooperation. Apple and Google, are probably seen as competitors, especially when it comes to mobile phones. But that wasn't always the case. When Apple launched the iPhone in 2007, they invited their counterpart Google as partners in this innovation. Apple iPhone had preinstalled Google Maps, and Google was the default search engine. Apple, working alongside Google as a preferred partner, would disrupt and dominate telecom, just as Microsoft had done working with Intel on personal computers just two decades before. Apple saw its relationship with Google as a preferential one, with complementary capabilities that would endure for some time. The two companies were collaborating to co-create value with smartphones, there was no competitive friction or combative tension between them. By 2009, Google launched the Android OS to compete with Apple's ISO. Unlike Apple, which made both hardware and software—OS for phones in a tightly integrated fashion—Google chose to make Android OS available to a network of hardware manufacturers to design and manufacture Android-compatible devices. Apple and Google had gone from being partners and co-creators of the iPhone smartphone to being competitors. Although Google's Android OS did not directly earn revenue, the Android ecosystem competes head on against Apple. In fact, Apple and Google are cooperative, competitive, and "coopetive", depending on the time and the industry.

The Apple-Google interconnection over the last decade provides useful pointers about the fundamentals of business relationships and the new frontier of co-creating value in digital ecosystems. The relationship among companies in evolving ecosystems are fluid and dynamic, as consumers want change and companies alter their set of capabilities to go after new pockets of value. As a result, no two companies are likely to be purely competitive or purely cooperative across time because their capabilities are neatly demarcated but complexly interdependent. And every company enters into many types of relationships. Managing such relationships to continually fine-tune your capabilities is key.

The idea of "coopetition" highlights how strategies for value creation and capture in the digital world differ from the classic principles refined in the Industrial Age. Pure competition is about dividing up the existing value pie, companies use their set of capabilities to win a greater share of the value. Cooperation is about enlarging and expanding the value pie by pooling the capabilities of several companies for the short term and also for the longer term. In contrast, "coopetition" is about both expanding the pie and ensuring that you get a fair share of the value. So at any given time, how do you determine who to cooperate with, who to compete against, and who to both cooperate with and compete against?

Adam M. Brandenburger and Barry J. Nalebuff wrote the influential book *Co-Opetition,* in which they say: "A player is your complementor if customers value your product more when they have the other player's product than when they have your product alone. A player is your competitor if customers value your product less when they have the other player's product than when they have your product alone. Since the same player can be both competitive and cooperative, these relationships need to be managed differently than if the complementor is not a competitor." They are quick to note: the relationship between competition and cooperation changes with time, especially while businesses transform in a digital era.

Co-creative capabilities with a digitalized ecosystem. "Coopetition" is central when traditional industries digitize because value is co-created with multiple companies across these traditional industries, and digital industries combine in ways to deliver new services to customers and capture value. Every company must coordinate the design and delivery of its products and services with platforms and solutions companies in ways to deliver the value that customers want. This is the essence of what's called: capability co-creation. According to your level of importance to partners and your partners' level of importance to you, there are four zones for you to plot the possible responses.

Transaction zone. If both you and your partners see little value in interacting either to cooperate or to compete, you could be experimenting on your own and the other players could be working on their own models of disruption. If there are any transactions between you and others, they are likely to be based on standard contracts as buyers and sellers. That is, you may be focused on digitalization trends and options, but you are thinking about what you could do by yourself with resources that could be acquired rather than through connecting with others. In this zone, there are really no specialized relationships, and you really are not spending much time thinking about "co-opetition."

Leader zone. It means you are in this zone as long as you are working from a position of strength. You are confidently defending your core strengths while adapting to digital ways of working. You may enter into a more preferential arrangement than standard contracts, but you are careful to protect your own advantage during this transformation. If a digital partner does not fit in with your selection criteria and frames of agreement, you might select a tech entrepreneur to work with you in this zone. Or you may involve industry incumbents to jointly explore new directions. In essence, you are leading your digital transformation shifts, and the other players are the supporting cast.

Follower zone. The other players are in the position of directing the transformation agenda, and you are more or less following their course of action, as defined by them. You may realize your lack of necessary competencies as your industry digitalizes and have no plausible option but to work with some of these disruptive interlopers, at least in the immediate term. You may enter a specialized relationship that gives you follow-up options, but in most cases, you are working together so that you learn quickly and understand the likely scale, scope, and speed of digitalization for your industry and business.

Co-creation zone. When you and your partners are working together to co-create value that neither of you could have done alone; you are both highly important to each other. Co-creation through "co-opetition" is both important and central to reinventing your business. Every "co-opetitive" relationship is different because all of them are multilayered, but what all of them share is a common understanding of what each party brings to the relationship and what their motives are. In this zone, both companies need to be more adaptive and master capabilities that resemble less joint planning and more dynamic improvisation with mutual respect and trust.

Capabilities to co-create. The key message from the four zones is about the inherent dynamics for the capability of co-creation across the various phases of transformation. In the Industrial Age, companies had well-defined roles, and relationships among companies were structured around a well-understood logic of what each party brought to the table. As you map your evolution along the four zones in the digital era, you will see that is no longer the case. Now, "co-opetition" is at the core, pulling and pushing different partners in relationships between incumbents, digital giants, and tech entrepreneurs. It is these interactions that influence the new ways in which businesses are developing their capabilities.

Structuring and managing your portfolio of relationships has always been important. Digitalization changes the context for your portfolio of relationships. The structure of your relationships changes over time as new digital technologies emerge and mature; the position of your relationships changes across the "co-opetition" as you make new choices, finance new investments, and set new priorities. The relative importance of your relationships changes as you and the other players make competitive moves.

Enumerate your capabilities to win in the digital future. The business capabilities you need to win in a world that is progressively digitalizing are different from the ones that you have mastered. You must select the distinctive capabilities. What is needed for you to become a master orchestrator across ecosystems? Are you arriving at your list of the key drivers of revenue and profits as your industry digitalizes and your company repositions within and across ecosystems? Decide what is core and what is to be co-creation. What can you make within your company that gives you a differentiated advantage and what can you co-create with others? The stronger your internal capabilities are, the more likely it is that you will attract stronger partners, and the easier it will be for you to absorb the digital capabilities within your organization.

Move selected key capabilities to the co-creation zone. You are looking at how capabilities and relationships evolve and assessing what you could do to move to the upper-right level, where you are co-creating capabilities that you may not need to do only by yourself. Manage the dynamic of core capabilities. The importance and impact of the digital initiatives change partly because your own priorities and preferences change but also because of competitive moves, maturing functionality, and commoditization. Also try to understand what might cause your partners to reposition your co-creative relationship; your partners may want to shift the relationship to a different zone as they pursue their own digital agenda.

Remember that digital business transformation lies at the nexus of scale, scope, and speed, which means that dynamic evolution is both implicit and central to your success. You might have been comfortable with the speed of shifts within your traditional industry, but the speed of digital transformation is faster and ferocious. The shape and structure of digital business ecosystems will change even faster in the future. Not only will this affect where you participate in ecosystems and when you orchestrate them, it will also dramatically change the nature of your relationships within these ecosystems. Knowing when to transact, when to lead, when to follow, and when to co-create is the key to your second winning move.

CHAPTER 6
MOVEMENT THREE: FROM TALENT TO INTELLIGENCE

MANAGING DIGITAL TRANSITION WITHIN AN ORGANIZATION

Driving change in a large, established organization is never easy, but it is even harder in the face of rapidly evolving technology and emerging business models that create huge uncertainties for the future. This is especially true for incumbent companies; they have assets that can't be ignored, while also having shareholders who demand profits. They have to strengthen their core and build for the future at the same time, a much harder task than starting from scratch.

Vision and a road map for the future. A vision and a sense of direction for the future are important when a company faces unprecedented challenges due to digital technology disruption and when employees and shareholders are uncertain about what the future holds for the company. Sometimes the future direction becomes clear when conditions limit options for current business practices, as *The New York Times* realized. Faced with a significant decline in revenue from classified advertising, the newspaper could not make up for it with online advertising due to low online ad rates and the dominance of Google and Facebook in online ads. The *Times* realized that its century old business model of relying heavily on advertising revenue was not tenable anymore. Cost cutting could help it survive in the short run, but it needed a vision and a path for the future. So, the company decided to focus on its other source of revenue, subscriptions, and plunged headlong into creating a pay wall, whereby readers had to pay for online news. In the fourth quarter of 2017, *The New York Times* added 157,000 new digital subscribers, a 41.8 percent increase compared with the end of the fourth quarter of 2016, ending the year 2017 with over 2.6 million digital subscribers. The company's total subscription revenue was almost double its advertising revenue, which is a major shift in its business model, which historically relied heavily on advertising.

Often there is a tendency to get seduced by the latest technology or the new business model of a hot startup. While incumbents should always learn from others, they should stay true to their core DNA and leverage the assets they have. The large installed base of General Electric machines provided the company a huge advantage in creating a platform for the Internet of Things, something that a startup would have a hard time mimicking. Walmart cannot become Amazon because it has thousands of stores that add to its fixed costs. However, it is exactly these thousands of stores that can be extremely valuable for reducing Walmart shipping costs for its online customers. In April 2017, Walmart announced a pickup discount for online customers who came to pick up online orders in stores.

Creating a road map for the future does not mean that the CEO has all the answers or knows exactly how the future will unfold. Instead, the goal is to provide broad direction, recognizing that the journey will never be linear and that the company will have to continuously evolve and shape its strategy within the broad guideline of its vision.

Navigating the turbulent transition period. Digital transition involves managing existing business and building for the future at the same time. It is like changing the engine of a plane while in flight. The plane is going to go down first before it goes up again, and that is a scary and uncertain time, when everyone in the organization starts questioning the company's strategy. *The New York Times* had no choice but to transition to a primarily digital newspaper and tilt its model toward subscriptions, even though print subscriptions and advertising still generated far more revenue than their digital counterparts. Imagine a scenario where in the long run the *Times* transforms itself from a print company to a fully digital company with no print version. Given declining print circulation and the high cost of printing and distribution, this is not a purely hypothetical scenario, in recent years several publications have decided to shift entirely to digital, such as *Newsweek* in 2012 and *InformationWeek* in 2013. If the *Times* moves to an all-digital future, its digital subscription and advertising revenues may be less than its current print subscription and ad revenue. However, being digital would also significantly reduce its production and distribution costs, which typically account for almost 50 percent of the total cost of a newspaper. In the end, the overall profit for the *Times* could be comparable to its current profit. While the future, all-digital profitability may be good due to lower costs of production and distribution, during the transition period the newspaper would be operating both the print and the digital business, which would increase rather than reduce its costs. In other words, the plane would go down before going back up.

Impact on internal operations. Digitalization often leads to substantial changes in the internal operations of a company, and every company should be ready for those changes to ensure a successful transition. As *The New York Times* developed its digital strategy, it also had to struggle with the issue of managing its print version. Who should get the privilege of breaking news, print or digital? If the digital group publishes the breaking stores, what should the print group publish the next day? The impact of new business models on the *Times'* operations was including product, price, promotion, distribution, customer management, sales force, and communication. What may seem like a small shift, from selling a package to selling a subscription service flawlessly on these internal changes, ensures a successful transition. All such changes should be supported by an appropriate organizational structure that leverages the assets and synergies of the firm instead of creating conflict between the old and the new organization.

Design, a loading dock for innovation. Starting an independent unit to spur innovation in a legacy company is like launching a speedboat to turn around a large ship, often the speedboat takes off but does little to change the course of the ship. Even though the speedboat was taking off, the mother ship was undergoing some pain.

Given the nature of industry, most tech entrepreneurs need large companies to get to scale. The incumbent can do things together and benefit from the innovation of the startups. Launching a speedboat that takes off as though it's in a different ocean might not be the best strategy. The best thing large companies can do is to make their ship accessible to startups and innovators. We could create a landing dock as a way for them to enter the ship. To create a landing dock, the company should ensure that everyone in the organization is completely aligned with the vision and strategy of the company and its operating rules.

More and more companies are beginning to realize that while it sounds intriguing and trendy to create a stand-alone—independent units full of energetic young entrepreneurs charged with outside-the-box thinking—these efforts often do not help in transforming a large legacy company. Creating a speedboat that is not tied, even with a long rope, to the mother ship does not help change the direction of the ship. And investing in startups without a clear idea of how they will end up on the landing dock of your ship is nothing more than playing a value capture game, not transforming your core business.

THE USE OF POWERFUL DIGITAL MACHINES TO CREATE NEW TALENT CAPABILITIES

What tasks could be automated, requiring minimal human intervention? What processes could be augmented with smart assistants? What jobs could be amplified with active interactions between humans and machines? These are the three key questions with which you should see the impact of the system of powerful technologies on an organization.

Automation of tasks. In the digital era, your business is at the nexus of scale, scope and speed, and that makes tasks much more complex than in the Industrial Age. The next generation of cloud robotics is no longer limited by the size of its memory or its own ability to compute data, it relies on data or code from a network to support its operation. No matter what your industry is, many of the tasks that you carry out could be fully automated, if not now, then in the very near future. About half of our traditional jobs are at risk, as computers take over automated tasks. But remember, using powerful computers is not simply about reducing the number of employees in your company, it's about competitive efficiency and effectiveness. If you automate faster than your competitors and if you automate a broader number of tasks, you will have a competitive advantage. Falling behind is bad enough when shifts in the marker are linear. It becomes notably more problematic when the shifts are exponential. The automation can mean freeing up employees for different work that brings greater value to customers.

Augmentation of processes. Although automation is a useful frame, what if powerful technologies could add value to, or augment, some of your tasks instead? Automation could help with tasks such as carrying out a first-order diagnostic for cancer or evaluating bids to acquire new television shows by looking at prediction based on past viewing habits of users. On the face of it, these look like tasks that could not be fully automated, but they can be thoroughly augmented by today's technology. In pilot projects, IBM Watson is already sifting through data and pulling out key numbers, computing ratios, and plotting—then mining—a comparative format that can serve as the first draft of a quarterly report. If you have recently read a financial news story on the *Forbes* website and did not scroll through to the end, you may have missed this note: Narrative science, through its proprietary artificial intelligence platform, transforms data into stories and insights. Simply put, a bot wrote that story. If you allowed powerful technologies to do this for you, what could the impact be on the people you employ? How many people might need to be retrained to work with such a machine? What skill mix might you need among your future employees?

Amplification of jobs. Where powerful machines really add value is by working with smart humans to expand the scale and scope of ideas, to amplify them, and this is where you should focus your company's attention. This is where your thinking cycles really help. Amplification depends on two principles: complementarity and singularity. Complementarity defines the areas in which machines are superior to humans and then creates governance rules and working conditions to bring out the best combined output and enhanced productivity. Singularity anticipates that intelligent computers will be capable of recursive self-improvement, or autonomously building ever smarter and more powerful machines. In other words, the first principle is about today and the second one is about tomorrow, and they act in progression. If you want to lead with amplification, where machines and humans accelerate the creation and capture of value, you need to create a climate that attracts the best future talent. It's a work environment in which employees learn at the frontier of how humans and machines work together; it's where software, data analytics, and algorithms are used to streamline decisions, where employee tasks cannot be done by machines today and they can apply their skills to work with machines to solve some of the profound challenges facing the world in energy, health care, space exploration, transportation, pollution, climate change, and so on.

Amplifying human talent with powerful smart machines. Figure out the number and type of people you need in the three clusters of automation, augmentation and amplification today, recognizing that this relative mix will change over the next decade. Your ability to win in the digital future may be more decided by your legacy human resource practices than by your legacy technology architecture.

First, recognize that computers are smart reasoning machines. Think about how computing and machine intelligence might allow you to solve complex problems and create new pockets of value for your customers. Think about how you might be able to augment your industry knowledge with powerful machines and orchestrate certain ecosystems and co-create with others within an ecosystem. How might reinventing your business with powerful machines at the core transform your organization and the logic of designing your work? Second, redesign how you think about work. Could machine intelligence—like IBM Watson—replace your job? What part of your job can machines do better than you? How should you redesign your job to take advantage of machine intelligence? Third, use powerful machines to create new value and new capabilities. What tasks could be automated, requiring minimal human development recommendations for video streaming subscribers, placing ads on search queries for everyone on a billion daily interactions across devices, including location and time? What processes could be augmented with smart assistants? What jobs could be amplified with active interactions between humans and machines?

Think about powerful technology such as machine learning, drones, robots, neural networks, big data analytics, cognitive systems, and algorithms. These technologies are just getting started, but development in one technology pushes development in others; they are interdependent. You have probably spent an inordinate amount of time thinking about organizational design in terms of structure, processes, roles, skills, and relationships using theories from organizational and social psychology. To succeed in the digital era, you need to start thinking like the digital giants and innovators that embrace computer science as the driving force for organization. Amplification that catalyzes humans with machines must be on your business agenda. It is time to develop your organizational system at the intersection of humans and machines and look at the critical skills that need to be mastered. What tasks carried on inside your organization, and by your key partners, could be automated with currently available and somewhat proven technologies? Could a joint initiative with a company such as IBM help free up your expert talent from mundane tasks to focus on more value-added areas of innovation and reinvention? By looking at other sectors and settings, you may be able to identify areas where and when amplification could become central to your business. Your initial classification of tasks and jobs is only a starting point. As you delve deeper into the frontier of automation, augmentation, and amplification at the intersection of machines and humans, and as new technologies further redefine these three zones, you need to reassess your classification and the talent pool you need.

DATA ANALYTICS, ARTIFICIAL INTELLIGENCE & TALENT MANAGEMENT

According to an IBM report in the early 2000s, we create 2.5 quintillion bytes of data every day. Ninety percent of data in the world today has been created in the last two years. This data comes from consumers' general activities such as web browsing, social media posts, and mobile usage, and it's increasing from sensors built into machines. One powerful approach for leveraging massive amounts of data is through machine learning and artificial intelligence. Today, AI is the force behind automation, and it creates fear and excitement at the same time. There is no doubt that AI will have a dramatic impact on the future of work and the skills needed to thrive in the digital era.

Impact of automation on jobs. The ability to collect data and train machines to analyze and learn is transforming every industry, and it is going to have a major impact on jobs. A 2017 study by McKinsey reported that while less than five percent of jobs have the potential for full automation, almost 30 percent of tasks in 60 percent of occupants could be computerized. Even highly skilled and well-educated lawyers and radiologists are under threat from automation. Automation in the 1960s and 1970s replaced blue-collar jobs in factories, but the type of automation driven by AI is likely to replace many white-collar jobs. Effectively, machines can more efficiently perform jobs that are routine, repetitive, and predictable—faster, and cheaper. The distinction is no longer between manual and cognitive skills, or blue-collar and white-collar work, but whether a job has large elements of repetition. Routine and repetitive parts of jobs will be automated, and people will need to retrain themselves for the non-repetitive aspects of the job. While some jobs will be eliminated, now jobs will be created that require new skills.

Not only are jobs changing, but the process by which firms recruit, develop, and manage talent is also undergoing dramatic change. Data and machine-learning algorithms will drive human resource decisions ranging from recruiting and training to evaluation and retention. Machines will not replace human judgment, but they will be major complementary assets to what we currently do to manage talent. The technology revolution is only going to accelerate in the future, and we better prepare and brace ourselves for it.

Recruiting. Technology is forcing firms to rethink who they hire and how they hire. Goldman Sachs has automated many parts of its IPO process and replaced most of its traders with software engineers who write algorithms. GE Digital has over 30,000 people with software and cloud-computing skills and those related to the Internet of Things. Gap comes up with new designs with mining data from Google Analytics and its own sales and customer databases. This increased focus on universal data across companies is leading them to hire people with skills in data analytics. Degrees in computer science and the ability to code are in great demand. The marketing function is also shifting, with greater emphasis on digital marketing skills.

How firms hire is also undergoing radical change. The traditional approach of interviewing is subjective and can introduce bias. Additionally, it is time consuming and limits the number of candidates a company can screen. Some companies are even beginning to question the reliability of traditional data points, such as college degrees and academic records, for spotting the right talent. Guy Halfteck, founder and CEO of Knack, a startup based in Silicon Valley, believes that by using mobile games and analytics this company can do as well as or better than companies that use the traditional interview process for recruiting consultants, financial analysts, surgeons, or people of just about any skill. Knack games create an incredible immersive and engaging digital experience. Playing a game involves up to 2,500 micro-behaviors per game, or about 250 micro-behaviors per minute. These include active and passive decisions, actions, reactions, learning, exploration, and more. Knack scores are computed from the patterns of how an individual plays the game, rather than how well they score on the game. From the raw data, automated analysis distills in-game markers of different behaviors on how the player processes information, how efficiently they attend to and use social cues like facial expressions of emotion, how they handle challenges, how they learn, how they adapt and change their behavior and thinking, and much more. These behavioral markers are articulated by a combination of machine learning and state-of-the-art behavioral science. Then Knack combines the behavioral markers to build and validate predictive models of psychology attributes, such as social intelligence, quantitative thinking, resilience, planning, and more. Each Knack score comes from one of these models, having developed 35 human behavioral attributes, and predicted real world behavior, such as job performance, leadership impact, ideation, learning success, and more. This data analytic and "gamification" approach to recruiting has won Knack many clients, including BCG, Citigroup, Nestle, IBM, Unilever, Walmart, to spot talent and broaden their pool of candidates. Typically, algorithms screen candidates in the early stages and face-to-face interviews happen only in the final phase. Unilever found that this approach was faster, more accurate, and less costly, and that it increased the reach of the company to a pool of candidates it had never interviewed before. Advances in analytics and AI have significantly improved the power and accuracy of people analytics. This is even more important in the gig economy, where many freelance workers are available for short periods of time for specific tasks and it would be too costly for a company to spend enormous resources in selecting these part-time employees.

Training and Development. Almost every company has online training courses and tools to help employees update their skills and learning. Using "gamification" that appeals to millennials, Adobe is able to send specific learning content to users based on their current level of knowledge and potential needs. The same technology can be adapted to customize training content for employees. Appical is helping companies onboard their young employees—a step in the journey to understand the specific needs and create customized training courses. Rapid changes in technology also warrant continuous learning. Senior executives in a company may be familiar with Snapchat and WeChat, which their target customers are immersed in, but it's their junior employees who really know how millennials use and engage with these technologies. Recognizing this gap, Unilever instituted a reverse mentoring program, in which a senior executive is paired with a young employee. The junior person helps the senior colleague in understanding the role that new technology plays in young consumers lives, and the senior executive mentors the junior partner about company strategy.

Performance Evaluation. Every Company uses some version of a performance evaluation system that includes 360-degree feedback and quarterly or annual reviews to set salary and bonuses and suggest improvements for the future. This traditional approach has three major limitations. First, it is very time consuming. Second, it is often ineffective; this biased approach leads neither to productivity improvement nor to employee engagement. Third, in batch mode employees often receive feedback only at the end of the year and not in the moment when it could help them improve their performance. Digital tools and technology are now allowing firms to test new and faster ways to assess employee performance. General Electric (GE) had an app PD@GE, which allowed workers to get real-time feedback from peers, subordinates, and bosses. GE is also now developing an app that uses past employee data to help leaders improve their succession planning and career coaching. Technology is not only useful in providing real-time feedback. The data-driven approach can also identify good performers without any bias inherent in ratings-driven evaluation. Royal Dutch Shell asked 1,400 people who had contributed ideas in the past to play Knack games. They then gave Knack information on how the ideas of three-quarters of those people had done in terms of seed funding or more. Using the game and performance data, Knack built a model, which was then used to predict the potential success of the remaining 25 percent of people. Without seeing the ideas, without meeting or interviewing the people who'd proposed them, without knowing their title or background or academic pedigree, Knack's algorithm had identified the people whose ideas had panned out. The top 10 percent of the idea generators, as predicted by Knack, were in fact those who'd gone furthest in the process.

Talent Retention. Retaining talented employees is a constant challenge for every organization. Often firms learn about imminent departures too late, when an executive has already secured another job and is ready to move on. This question has led to the development of new tools that are able to predict the likelihood of an employee leaving a firm. GE is currently testing such an application, to predict six months in advance if an employee is likely to leave, so that the company can design an appropriate intervention before it is too late. Scores of academics are using their skills in customer-churn modeling to develop machine-learning algorithms for predicting employee turnover.

PART THREE

LEADING DIGITAL

CHAPTER 7
NEW PERSPECTIVES

Gandhi said: "You must be the change you wish to see in the world." It is so critical that you know who you are, where you are, and have the vision to see the future within your industry, while leading your business in that direction.

A NEW DIGITAL MINDSET FOR PRINCIPLE AND PRACTICE

Every company faces its own unique digital inflection point. Past business models are showing signs of ineffectiveness. New directions and options with digital technologies are daunting. Savvy digital upstarts are intimidating. However, the choices made at such points in time matter because they define the subsequent avenues and follow-up options—just as Kodak, Nokia, Blackberry and Sony either failed to see, or simply could not adapt, when faced with these inflection points.

Digitalization is an ongoing evolution that influences how important problems are solved in digital economies, industries, and society—how companies are created and re-created, and how organizations are designed and re-designed to deliver superior value to consumers. Just as the industrial revolution was neither a single point in time or a single technology in a single industry or geography, digitalization will have far-reaching global impacts in the 21st century and beyond. The bottom line is that it's not too late to step up to the digital transformation space, but the rate of change will be rapid, and the risk of inaction will soon be more expensive than the risk of experimenting and getting started.

Clarify your long-term objective. Look at the initiatives that your organization has undertaken. How comprehensive are they? How could they be refined? The digital agenda could help you match your long-term goals with your current capabilities to know what your next moves should be. For example, the Accor hotel group in France has instituted digital hospitality programs that target clients, employees, and partners with systems and data. What follow-up actions would be next? Would amplifying human talent with powerful machines allow the company to co-create new capabilities and provide customers with new pockets of value? Or might orchestrating one or more of the hospitality ecosystems allow the company to increase its internal efficiency and customer service, thereby expanding the scope of the program? Both are valid directions, but one will probably be a more logical extension of the company's position at the intersection between the three phases, the three sets of players, and the three winning moves.

Similarly, we've seen that Siemens has announced its next47 unit to focus on disruptive ideas to accelerate new digital technologies. What specific actions might make the most sense for Siemens' long-term strategic innovation? Should it seek to co-create in the leader zone with specific tech entrepreneurs, or digital giants, to differentiate itself from other incumbents? Might it be better for the company to develop its digital systems so that it can take the company in new directions? Again, both are possible directions, but one is likely to be a better fit for Siemens in the current environment. The point is that you can be in any company, in any industry, nearly anywhere in the world, and you will find ways to apply the principle and practice of digital transformation. Your success lies in doing this more adeptly than your competitors and your collaborators.

Think about new winning moves. We should not be blind to the possibility that another winning move might emerge, perhaps a move focused on an area such as security, privacy, or identity. These three areas are becoming increasingly important and deserve extensive discussions of their own. A winning move must truly challenge the direction of your digital transformation and offer you new insights for inventing your business model at the scale, scope, and speed you need to win. The three phases of digital transformation are enduring and evolving. New winning moves are important business principles that could be supported by additional attention to specific technical areas.

There is no one universal principle of digital business and there is no one best way to make the transformation. However, the digitalization of your business is the most important item on your company's strategic agenda. You may have started along the transformation journey in different parts of your organization and now need to coordinate these piecemeal initiatives.

NEW TECHNOLOGY DRIVEN

Nevertheless, technology is always the key driver for business digital transformation, no matter what kind of industry with which you are involved. The confluence of elastic cloud computing, big data, AI, and IOT highly drives digital transformation. Companies that harness these technologies and transform into vibrant, dynamic digital enterprises will thrive. Those that do not will become irrelevant and cease to exist. If the reality sounds harsh, that's because it is. While the cost of failing to adapt is perilous, the future has never looked brighter for large companies embracing digital transformation.

The first reason that those giant incumbents in the business world can take advantage of digitalization is Metcalfe's law: Networks grow in value as the participants increase. Large corporations stand to benefit from a similar paradigm regarding data. If properly used, the value of enterprise data also increases exponentially with scale. Large companies tend to have dramatically more data than the upstart competitors seeking to supplant them and can collect data considerably faster. If incumbent organizations can digitally transform, they will establish the most data, which is an asymmetric advantage that could dissuade competitors from easily entering their industries.

The second reason lies in the fact that giant incumbents are well positioned to exploit digital transformation because they typically have access to substantial capital. Digital transformation offers highly attractive investment opportunities. One such opportunity is hiring large numbers of top-flight data scientists and engineers. Another is investing in digital transformation. As it turns out, these two factors—data moats and access to capital—work synergistically. Large companies with proprietary data, the right technologies in place, and the capital to recruit top talent will find themselves in almost unprecedented positions.

Over the past few decades, the information technology revolution went from mainframe computing to minicomputers, to personal computing, to Internet computing, and on to handheld computing. The software industry has transitioned from custom application software based on Multiple Virtual Storage (MVS), Virtual Storage Access Mode (VSAM), and Index Sequential Access Mode (ISAM), to applications developed on a relational database foundation, to enterprise application software, to SaaS, to handheld computing, and now to the Artificial Intelligence-enabled enterprise. The Internet and the iPhone change everything. Each of these transitions represented a replacement market for its predecessor. Each delivered dramatic benefits in productivity. Each offered organizations the opportunity to gain sustainable competitive advantage.

We have seen that each of the key technologies driving digital transformation, including elastic cloud computing, big data, AI, and IOT, presents powerful new capabilities and possibilities. But they also create significant new challenges and complexities for organizations, particularly in pulling them together into a cohesive technology platform. In reality, the technical requirements to enable a complete, next-generation enterprise ecosystem that brings together cloud computing, big data, AI, and IOT are extensive. They include core requirements such as data aggregation, multi-cloud computing, edge computing, platform services, enterprise semantic models, enterprise microservices, enterprise data security, system simulation using AI and dynamic optimization algorithms, open platform, as well as a common platform for collaborative development.

The current step-by-step function in information technology has several unique requirements that create the need for an entirely new software technology stack. The requirements of this stack, to develop and operate an effective enterprise AI or IOT application, are daunting. It is necessary to aggregate data across thousands of enterprise information systems, suppliers, distributors, markets, products in customer use, and sensor networks, to provide a near-real-time view of the extended enterprise. Data velocities in this new digital world are quite dramatic, requiring the ability to ingest and aggregate data from hundreds of millions of endpoints at very high frequency, sometimes exceeding 1,000 Hz cycles. The data needs to be processed at the rate it arrives, in a highly secure and resilient system that addresses persistence, event processing, machine learning, and visualization. This requires massive, horizontally scalable elastic-distributed processing capability offered only by modern cloud platforms and supercomputer systems. The resultant data persistence requirements are staggering in both scale and form. These data sets rapidly aggregate in hundreds of petabytes, even

Exabytes, and each data type needs to be stored in an appropriate database capable of handling these massive volumes at high frequency. Relational databases, key-value stores, graph databases, distributed file systems, blobs, all are necessary, requiring the data to be organized and linked across these divergent technologies.

Reference AI software platform. The problems that have to be addressed to solve the AI or IOT computing problem are nontrivial. Massively parallel elastic computing and storage capacity are prerequisites. These services are being provided today at increasingly low cost by Microsoft Azure, AWS, IBM, and others. This is a huge breakthrough in computing. The elastic cloud changes everything. In addition to the cloud, there is a multiplicity of data services necessary to develop, provision, and operate applications of this nature. Here is a list of some of the data-service requirements for digitalization of giant incumbents in the business world.

Data integration. This problem has haunted the computing industry for decades. One prerequisite to machine learning and AI at industrial scale is the availability of a unified, federated image of all the data contained in the multitude of information systems such as ERP, CRM, HR, MRP, as well as Sensor IOT networks such as SIM chips, smart meters, programmable logic arrays, machine telemetry, and bioinformatics. Also, relevant data such as weather, terrain, satellite imagery, social media, biometrics, trade data, pricing, and market data are critical for data integration.

Data persistence. The data aggregated and processed in these systems includes every type of structured and unstructured data imaginable: personal identity information, census data, images, text, video, telemetry, and voice, network topologies. There is no one size fits all database that is optimized for all these data types. This results in the need for multiplicity of database technologies, including but not limited to, relational, key value stores, distributed file systems, graph data bases, and blobs.

Platform service. A myriad of sophisticated platform services is necessary for any enterprise AI or IOT application. Examples include access control, data encryption in motion, encryption at rest, ETL (Extract Transform and Load), queuing, pipeline management, auto-scaling, multi-tenancy, authentication, authorization, cyber-security, time-series services, normalization, data privacy, etc.

Analytics processing. The volumes and velocity of data acquisition in such systems are blinding and these types of data and analytics requirements are highly divergent, requiring a range of analytics processing services. These include continuous analytics processing, batch processing, stream processing, and recursive processing. In addition, data visualization tools are also important during analytics processing and presenting. Any viable AI architecture needs to enable a rich and varied set of data visualization tools including Excel, Tableau, Spitfire, Oracle BI, Business objectives, Domo, Apteryx, and others.

Machine learning services. The whole point of these systems is to enable data scientists to develop and deepen machine learning models. There is a range of tools necessary to enable that, including Python, DIGITS, and SCALA. Increasingly important is an extensive curating of machine learning libraries such as TENSORFLOW, CAFFE, Torch, Amazon machine learning, and AZURML Your platform needs to support them all.

Developer tools and UI frameworks. Your IT development and data science community, each have adopted and become comfortable with a set of application development frameworks and user interface tools, including the VI, visual studio, react, etc. In addition, being open, extensible, and future proof is also important. It is difficult to describe the blinding pace of software and algorithm innovation in the system described above. All the techniques used today will be obsolete in 5 to 10 years. Your architecture needs to provide the capability to replace any components with their next-generation improvements, and it needs to enable the incorporation of any new open source or proprietary software innovations without adversely affecting the functionality or performance of any of your existing applications. This is a level-zero requirement.

Cloud vendor tools. An alternative to the open-source cluster is to attempt to assemble the various services and micro-services offered by the cloud service providers working the seamless and cohesive enterprise AI platform. Leading vendors like Amazon Web Services (AWS) are developing increasingly useful services and micro-services that in many cases replicate the functionality of the open-source providers and offer new and unique functionality. The advent of this approach over open source is that these products are developed, tested, and quality assured by highly professional enterprise engineering organizations. Moreover, these services were generally designed and developed with the specific purpose that they would work together and interact in a common system. The same points hold true for Google, Azure, and IBM.

In a word, when a new and more powerful technology becomes available, your applications keep on ticking but now with greater performance, precision, and economic benefit. All the developing new technologies are fundamentally driven, but requirements for business digital transformation.

NEW DIGITAL INTERNET ECOSYSTEM COMMITMENT

Based on taking advantage of the advanced digital technologies such as big data analytics, cloud computing, artificial intelligence, IOT, digital transformations are not only mainly focused on the front line and end consumption of economy, society and the world, but also the inside knowledge of operating a supply chain in various industries. It shows that the path of digital transformation is in the flow back to C2B2B, which is the progressive path from front-end consumers, back to retail business, to distribution, to brand business, to manufacturing, and to raw materials and component parts supplying. This is termed as from demand side (business to consumer) to supply side (business-to-business). This economic phenomenon in digital transformation is called Industrial Internet 4.0.

The revolutions of Industrial Internet 4.0. This development of Industrial Internet 4.0 makes magnificent changes in a traditional industry. The key intentions of Industrial Internet 4.0 are to enhance industrial efficiency and create innovative value. With the application of digital technologies, the traditional industry businesses can improve in four areas. The first one is to optimize the industrial supply-chain. The second is to reduce the production process. The third is to improve the operating system. And the last, but not the least, is to utilize Internet applications. All the efforts on Industrial Internet digital transformation will lead to the result of building up the hyper-connectivity and convergence both of supply and demand side, in order to achieve real-time responsive production and collaborative operation.

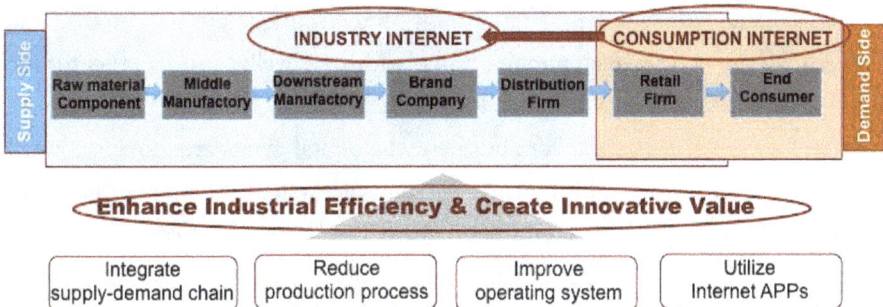

Industrial Internet 4.0 serves to break through the information barriers among different functional units of organization—internal and external—with various business companies and industries in a digital ecosystem. It shapes up the decentralized business platform and dispersed business model, which is a hyper-connection ecosystem incorporating both supply side and demand side. This helps to enhance the information exchange and interactive response in innovative methodology, and critical flows—the flow of row material and products, the flow of production and supply chain, and the flow of capital and finance—are seamlessly connecting within the entire integrated business and industry ecosystem.

The growth consumption Internet is based on the number of users and their attention. It means increased Internet traffic is the most important subject for scope, scale expansion, and speedy growth of companies no matter if it is the consumption Internet giants or a startup unicorn. With the ceiling of growth with Internet traffic, business in the consumption Internet must change their mindset to look for new opportunities in other areas. It makes a lot of sense that the companies try to extend their business scope and scale toward new areas of the upper stream among the entire industry chain. New technologies such as the Internet of Things, data analytics, and artificial intelligence are applied in various levels to provide more solutions for efficiency improvement and value added for the achievement of business objectives.

Meanwhile, innovation in digital technologies also empowers the traditional industrial incumbents with new opportunities to succeed in a digital transformation, interconnecting within the digital economic ecosystem, and collaborating while competing with other digital giants and entrepreneurs, for the solutions of optimizing production process, increasing operation efficiency, reducing production cost, as well as managing consumer's perceptions.

In the age of the consumption Internet, digital giants such as Amazon, Apple, or Alibaba, and Meituan, have already built up their unique business model. The industry infrastructures, technologies, and platforms as well, come together for the development of their own business kingdom. While in Industrial Internet 4.0, starting from their successful core business, the digital giants are moving toward an upstream supply chain in the entire business ecosystem, extending to achieve breakthrough growth. For example, Alibaba's core business is e-commerce; they spread both online and offline business with online supply chain production and offline new retail business. Meituan's core business is delivery; it extends to other businesses focusing on targeting key consumers based on all the consumer data gathered from all the interactive transactions via the core business.

Digital transformation in the Industrial Internet 4.0. Industrial Internet 4.0 is more than technology. It is also about taking advantage of innovative technologies to achieve digital transformation, including rebuilding industrial research and design, production processes, organizational operations, and marketing & sales functions, to achieve business revenue and profit along with the advanced networking connection through new technologies, new platforms, and a new management mindset.

In the Industrial Internet 4.0, the role of the research and development department is no longer taking responsibility for research, design, production, and evaluation. It is focusing on the de-centralized variety of consumer needs, working for R&D outsourcing, design platform, and open-to-publish evaluation of product offerings.

In the Industrial Internet 4.0, the traditional production system, which includes the process of material preparation and quality control, is totally adapted into different systems that emphasize personal customization, agile production line, database analysis, and seamless service.

In the Industrial Internet 4.0, the task of marketing is not only about market analysis and customer study, communication campaign design and execution, distribution dealer management, after-sales service, and relationship maintenance, but also data intelligence, dynamic pricing, RIT (Real in Time) bidding platform, and data-sharing alliances.

In the Industrial Internet 4.0, the organizational operating process—beyond establishing objectives and strategy—sets up the functional and project-based structure and attains the performance measurement. The process mainly includes necessary solutions: synergizing information, and accurate resource allocating.

The hype-connected ecosystem of Industrial Internet 4.0. The Industrial Internet 4.0 is hype-connected within an entire digital ecosystem that is involving many more shareholders. It is a much longer chain with more complicated situations than the core manufacturing. It also includes the different kinds of resources in the entire business chain, such as research, design, production, distribution, sales service partners, communication agencies, and in the end, consumers as well. In fact, the entire Industrial Internet 4.0 ecosystem is built with three tiers. The first tier is infrastructure, which is about the IOT, mobile communicating, and cloud computing. The second tier is service, which is about supply-chain financing, logistics, technology support, talent management, and marketing. The third tier is industry, which is about the value chain creation of industry and its accompanying operating system, from upstream supply chain to downstream end users.

In traditional industry, the scope and scale extension are based on an assembly line that handles buildup. Business process models started from supply chains of production-inventory-marketing-consumption. This model means long production periods, large inventory, a lack of product personality, and it cannot meet the needs of new generation consumers in the marketplace. The good news is that Industrial Internet 4.0 can change the production-consumption model by taking advantage of digital technology, connecting the demand-supply and supply chain to build up an agile, interactive response business model, and to bring a seamless integration of value chain for a great end-consumer experience. Meanwhile, in Industrial Internet 4.0, manufacturing, as a service of the industry's professional business platform, can share the production resources and construction. For example, 3D print technology can be applied on the MaaS platform to connect producer and consumer, to accomplish fast production while maintaining a low-cost manufacturing service. It is a vision that MaaS has the capability to build up cooperating network connecting data with production resources of capital, equipment and so on.

From one side, the advanced technologies, including big data, cloud computing, artificial intelligence, the IOT, empower the Industrial Internet 4.0. On the flip side, Industrial Internet 4.0 also pushes advanced technologies to improve application and adaptation, to business operation and growth. It is the convergence of digital technologies and the Industrial Internet within a digital economic ecosystem that establishes a powerful new business world with sustainable development.

CHAPTER 8
NEW CONSUMERS

Growth is a key priority for every business, and acquiring new customers is a major driver of growth. Digital and social marketing tools provide new and innovative ways to spur this growth. However, acquisition costs and profitability vary widely between customers and channels. So which customers should be acquired, how to engage them, and what is the value management of the lifetime consumer? These are the fundamental questions, especially in a digital economic era, and also the data equity advantages to be taken during business digital transformation.

ACQUIRING NEW CONSUMERS

The kinds of customers to acquire. In 2016, Chase introduced its new Reserve Card with a very attractive offer: a signing bonus of 100,000 points, $300 in travel credit, triple points earned on all travel and dining expenses, and a value of 1.5 times applied to points redeemed on travel. This offer generated so much enthusiasm among consumers that within a month of the card's introduction Chase ran out. Chase said that the company exceeded its annual target of customers in less than two weeks. This generous offer was expensive for the bank. In December 2016, the new Reserve Card would reduce the bank's fourth-quarter profits by as much as $300 million. It would take the bank five and a half years to break even on its promotional investment in the card. With the card's high annual fee, Chase was clearly aiming for the affluent customers who have historically been the prime target of the American Express Platinum card. Yet surprisingly, the majority of people who signed up for the card were millennial. Chase justified the significant acquisition of these customers because millennials make up the majority of their new deposit accounts today, and their wealth is expected to grow at the fastest rate of all generations over the next 15 years. However, whether or not acquiring these customers was the right decision for Chase depends on how many of these new customers will stay with Chase after the first year, especially when competing firms are also making attractive offers to them. The question is whether the type of consumer this attracts leads to a less profitable card product in the long run. In February 2018, to address analysts' concern, Chase provided an update on the new Reserve customers. It stated that the average income of new cardholders is $180,000. Their annual spending on the card is $39,000, and their retention rate is over 90 percent. Chase also started a pilot program in 2018 to convert many of its Reserve Card customers into Chase private client and mortgage customers.

The Chase example illustrates that the number of acquired customers, or the acquisition cost, does not provide sufficient information to evaluate a customer acquisition strategy. It is critical to know customer's spending and retention rates to estimate their long-term value. Yet most companies track a host of short-term metrics to assess their marketing campaigns, such as impressions, number of clicks, click-through rate, conversion rate, and customer acquisition cost.

According to the familiar 80-20 rule, 20 percent of the customers provide 80 percent of the revenue. However, research shows that if we focus on profitability instead of revenues, the rule would be 200-20, where 20 percent of the customers provide almost 200 percent of the profit. How is that possible? It is because the remaining 80 percent of customers actually destroy profitability. In other words, a company's profit would soar if it were to jettison the bottom 80 percent of its customers. Of course, some of these unprofitable customers may be important for other, strategic reasons, but this analysis forces management to articulate the reasons for retaining these unprofitable customers. These findings underscore the importance of acquiring and retaining the right customers, those who are likely to be profitable in the long run. It also suggests that simple metrics like total number of customers, or overall market share, may be misleading. A company with a large market share may be saddled with a high proportion of unprofitable customers.

Even if a firm recognizes that it should acquire customers based on their expected long-term profitability, it faces several challenges in implementing this idea. First, most companies are organized by product or business units that mask variation in the profitability of customers within that product or business unit.

Second, to measure customer profitability, a firm needs to adopt activity-based expenses to allocate costs to each customer or customer segment. While this task may seem tedious, advances in cost accounting, data analytics, and technology-based solutions are making it easier to achieve this objective. Third, firms need to keep track of cohorts of customers, to understand their long-term profitability and to allocate resources accordingly. Aggregating customers in a single database—regardless of where they come from—makes it difficult to assess the effectiveness of your customer acquisition program.

How to acquire customers. The customer acquisition process must begin with a deep understanding of a consumer's decision journey or path to purchase. Procter & Gamble described two moments of truth—the best brands consistently win with these moments of truth. The first moment occurs at the store shelf, when a consumer decides whether to buy one brand or another. The second occurs at home, when they use the brand, and are delighted, or aren't. Introduction of electronic scanners in stores in the 1980s—and the rise of e-commerce in the last two decades—have made it possible for firms to know the first moment of truth by tracking the sales of their products. Although P&G could measure sales, the firm had no knowledge of how people actually consumed and experienced products, the second moment of truth. Recent developments in technology are making it possible to track product consumption. For example, Nestle could introduce chips into Nespresso machines to know consumers coffee-consumption behavior and automatically order Nespresso capsules when consumers' supplies got below a certain threshold. The battle for consumers begins long before they buy or experience a product. It starts when they open their laptops or tap on their mobile phones to search for a product. In 2011, Google coined the term Zero Moment of Truth (ZMOT) to reflect the importance of this period of online searching before consumers show up in a store or make an online purchase. In a Google study, 84 percent of shoppers claimed that ZMOT shaped their decision of which brand to buy. Using consumers search data, Google created heat maps to visualize how and when consumers actively searched for a product. For example, a consumer's search for a new automobile is typically most active two or three months before actual purchase. In addition to doing Google searches, consumers read product reviews on Amazon, hotel reviews on Tripadvisor, restaurant reviews on Yelp, and movie reviews on Rotten Tomatoes before making any decision to buy a product. The rise of social media and the increasing use of reviews by consumers have made the third moment of truth a critical factor. This is when loyal fans for a product become passionate advocates for it on social media and consumer-review sites. A vital part of any successful customer-acquisition strategy is to understand these moments of truth and ensuring that a brand is well represented at each stage.

From search to purchase. The consumer journey from the ZMOT to the Final Moment of Truth (FMOT) typically involves four stages: awareness, consideration, evaluation, and purchase. For a long time, this process was believed to be linear. A consumer may be aware of eight brands of smartphones, but she may consider just three brands seriously, brands whose various product features and prices she evaluates more thoroughly, leading ultimately to her decision to buy a specific brand. Recent studies by McKinsey and others have shown that consumers' search processes may not be linear. A consumer looking to buy a car may start with only BMW and Mercedes, but in the course of searching, may come to consider other brands in their deliberations. McKinsey has found that for automobiles consumers have, on average, 3.8 brands in their initial consideration set, but they add 2.2 additional brands during their search. Therefore, brands have an opportunity to influence consumers' decisions during the search process. A firm can provide information and influence consumers in three ways: through paid media, which would involve search engine marketing (SEM), and ads on TV, radio, newspaper and magazines, through owned media, which would include the company website leveraged with search engine optimization (SEO) to ensure that the site ranks high in organic online searches, and through earned media, whereby consumers learn about a product from the reviews and opinions of other consumers on social media.

Personalization and retargeting. Digital technology and rich data allow firms to personalize ads to consumers based on their individual interests, web-browsing behavior, past purchases, and the context of the site they are visiting. A novel study by MIT went one step further by morphing banner ads to match consumer's cognitive styles. For example, some consumers like to read text while others are more visual. If an advertiser can define which style any particular Internet user possesses, it can enhance the potential effectiveness of an ad by matching the ad to a given consumer's cognitive style. The MIT team surveyed a sample of consumers to learn their cognitive style and then traced their web-browsing behavior to link the two. In practice, one can observe only consumers web-browsing behavior, not their cognitive styles. But by using Bayesian models and estimates from the sample consumers, the MIT team could infer consumers' respective cognitive styles and service personalized ads in real time. The MIT researchers tested this approach in a large-scale field experiment in which more than 100,00 consumers viewed over 450,000 banner ads on CNET.com. Morphing doubled the click-through rates of the ads. In the follow-up experiment for automobiles, the researchers further demonstrated not only that ad click-through rate improved with morphing but also that brand consideration and purchase intentions also jumped significantly. In another follow-up study, they developed an algorithm to morph the entire website in real time. Another commonly used approach to improve ad effectiveness is called retargeting, in which ads are shown to consumers who previously visited a firm's site but did not buy. Several studies have shown the effectiveness of retargeting, including a recent large-scale field experiment for an online sports company. Using Google's display network of two million websites, the experiment showed that retargeting increased website visits by 17 percent, transactions by 12 percent, and sales by 11 percent. But how specific should retargeting be? A specific of dynamic retargeting shows consumers an ad is for the exact product that they searched for previously, whereas a generic retargeting may simply show an ad is just for running shoes. A field experiment for an online travel company, in which almost 80,000 consumers participated, revealed that specific or dynamic retargeting surprisingly does worse than generic retargeting. The result of the study suggests that many consumers, who often believe their data is being collected, do not have well-formed preferences, especially early in their purchase process, which make specific targeting less effective.

Social media and "Viral Marketing." Social media has gained a lot of attention from marketers since it allows a message to be amplified without additional cost to a firm. Experts also think that social media is more effective than traditional ads since consumers believe the opinions of other buyers. The idea of "creating an online campaign and making it go viral" is seductive. Without spending much money, you create a video ad that is shared and viewed by millions of consumers. Marketers are quick to quantify the value of this earned media in terms of the dollars saved that would otherwise have been required to reach the same number of consumers through a paid media campaign. But can you really engineer going viral? The average YouTube video generates fewer than 10,000 views, and only a tiny fraction of YouTube videos have more than one million views. The notion of viral marketing comes from epidemiology and involves a single person infected with a communicable disease spreading it to a large population. For a disease to reach epidemic proportions it needs a reproduction rate of greater than one, so that each person who gets the disease will spread it to more than one person. Otherwise, the disease dies down quickly. A 2012 study examined the spread of millions of messages on Twitter and Yahoo and found that more than 90 percent of the messages did not diffuse at all. About four percent of the messages were shared only once, and less than one percent were shared more than seven times. To address the challenge of creating a viral video, BuzzFeed proposed the idea of big-seed marketing. It is suggested to seed the message with a large number of people in the hope of amplifying it to a larger audience even if the reproduction or sharing rate is less than one. Recognizing the difference between viral and amplification, BuzzFeed suggests companies to aim amplifying messages through native advertising. Amplification depends not just on creating intriguing stories with humor and catchy titles, but also on the medium and the authenticity of the message. The brand marketers should understand this: you can trick people into clicking, but you can't trick them into sharing. Everything that performs well is based on a real insight, something that's actually true about the brand. Who should seed the information? Most marketing executives believe that social media influencers are the best bet. However, I realize it is hard to find reliable influencers. Even using people with a large following on social media does not guarantee success. In 2009, a year before the launch of its new Fiesta, Ford recruited a hundred influential social media personalities to promote the car through blogs, videos and photos. The list included someone with 115 million views at that time. However, this turned out to be one of the least effective agents for Ford. Although under some circumstances, the most influential users are also the most cost-effective, under a wide range of plausible assumptions, the most cost-effective performance can be realized using ordinary influencers individuals who exert average, or even less than average influence. Collectively, these studies suggest that trying to go viral by seeding a message with a handful of social media influencers is unlikely to be successful and cost-effective. Instead, it is better to seed the content with a large number of ordinary users. In fact, many social campaigns gain traction only after being picked up by the mainstream media.

Acquisition through Offers. Companies in almost every industry typically offer significant discounts to attract newcomers. How do these offer-based acquisition methods compare with word of mouth or referral programs? Two studies provide some insights. Using data from a web-hosting company, one study found that while marketing-induced customers add more short-term value, customers acquired through word of mouth add nearly twice as much long-term value to the firm. Although short-term discounts may provide a quick win in acquiring new customers, word of mouth and referral programs are more effective in the long run. In practice, a company faces customers with varying price sensitivities. Some customers are more price driven and won't buy unless they are offered discounts. To separate the price-sensitive customers from others, brick-and-mortar retailers have long followed the practice of creating search friction, placing the discounted items in the back of the store or in a separate outlet store. In contrast, online retailers try to reduce search friction and improve user experience for every customer. This results in either offering lower prices to everyone, or charging relatively higher prices at the risk of losing price-sensitive consumers. Now the e-commerce companies could deliberately create search friction for discounted products that only price-sensitive customers would seek. For a group of consumers who are not price-sensitive, they eliminated discount markers, which resulted in the consumers in this group buying fewer discounted products—and at lower average discount—without any impact on their conversion rate, leading to a significant increase in the companies' online retail profits. Recognizing that not all customers are equally price sensitive or equally valuable, firms may want to consider deliberately making it harder for consumers to find discounted items.

ENGAGE NEW CONSUMERS

The right message, to the right customer, at the right time, marketing executives around the world are convinced that this goal is now becoming a reality, thanks to their ability to track every single click of a consumer on the web, know her location at any time through her mobile device, and have a deep understanding of her interests and activities through social media and Facebook (Meta). Sophisticated approaches such as programmatic media buying, real-time marketing, data mining, geo-targeting, and retargeting have given marketing experts new confidence in their ability to achieve this goal. Yet most consumers find ads annoying. A 2016 survey found that 90 percent of consumers skip pre-roll video ads or 84 percent of them skip mid-roll video. In spite of all the developments in technology and the rhetoric about engaging consumers, marketers have failed to account for the consumer's perspective. Every brand in the world wants to engage with the consumer, but why would a consumer want to engage with a bar of soap, a can of soda, or a bottle of beer?

From storytelling to story-making. It is very impressive that MasterCard's iconic "priceless" ad campaign first launched in 1997. This campaign showed vignettes of human interactions that concluded with the lines "There are some things in life money can't buy, for everything else, there's MasterCard". The campaign was so successful that in the next fifteen years it entered the vernacular in many countries. Yet it was time for a change when it comes to rethinking how to engage consumers. To drive this change, the marketing team decided to position the brand as "connecting people to priceless possibilities." To bring this to life, they created a digital engine that leveraged digital and social media. The digital engine is a seven-step process based on insights gleaned from data and real-time optimization. *1. Emotional spark.* The first step is to create an emotional connection with consumers. Using data to understand consumers' key passion points, MasterCard builds videos and other creatives to ignite this spark and give consumers a reason to engage. For example, a few weeks before the 2014 New Year's Eve, MasterCard produced a video in which the actor Hugh Jackman announced a promotion encouraging consumers to submit a story about someone who deeply mattered to them. The authors of the winning submissions would, on New Year's Eve, be flown anywhere in the world to reunite with those both distant and dear. MasterCard envisioned each reunion as a priceless surprise. *2. Engagement.* By using data to identify the right audience, MasterCard targets that audience with a spark video through Facebook (Meta) and other social media. The goal is to encourage consumers to share their stories. To continue this excitement, MasterCard often produces video, in the Hugh Jackman case, the company showed Jackman surprising his own mentor in New York. *3. Offers.* With the goal of helping its partner banks and merchants drive their business, and in turn MasterCard's own revenue, the company identifies the best offers to match consumers' interest. In the New Year's Eve campaign, MasterCard's Asia-Pacific team partnered with Singapore's resorts with an attractive offer. *4. Real-time Optimization.* At any point in time MasterCard may have several offers. The one that should be highlighted and promoted is determined by AB testing and real-time optimization of offers, themes, and budget allocation. *5. Amplification.* Real-time testing gives MasterCard confidence about the potential success of these offers, and it also encourages its bank and merchant partners to co-market and co-fund these campaigns. This process amplifies both MasterCard's budget and the impact of the program. *6. Network effects.* A few weeks after consumers have submitted their stories, MasterCard selects winners, produces video of them surprising their friends and families, and posts these videos on social media to encourage sharing. *7. Incremental transactions.* These programs translate into incremental business for banks that issue cards, for merchants where consumers spend money, and for MasterCard, which gets a portion of every transaction. It is true that no one is listening to our story, no matter how compelling it is. A broader message that is emotionally engaging allows for a two-way conversation. It shifts from storytelling to story-making and keeps consistent with the brand value, while also creating engagement that drives business.

Moment based marketing. Advertisers are obsessed with knowing consumers, their demographics and interests, their network of friends, what they post on Instagram or Pinterest, what they write on social media. The more we know about the consumer, the more targeted and relevant the ad will be. But consumers are complex and multidimensional. Their mindset and receptivity to any message varies greatly depending on the context. Marketing experts have always focused on macro-moments such as the Thanksgiving holiday or TV prime time. However, we now live in the mobile era, when on average, consumers check their smartphones 150 times a day. More than 87 percent of consumers have their phones by their side's day and night, and 68 percent admit to checking their phone within fifteen minutes of waking up in the morning. Unlike television ads that aired at 8 PM, regardless of whether or not that was the right time for a consumer, we now have the ability to wait for the right moment before sending a message to a consumer's mobile phone. This is the era of micro-moments, when messages need to arrive at the proper time, in the proper context. For example, at Sephora, shopping in stores can be overwhelming for consumers because they face a large number of choices. It is not uncommon for consumers in these situations to pull out their smartphones in order to find product ratings and consumer reviews. This provides a great opportunity for retail customers to leverage their phone as a shopping assistant when they are standing in the store. To help shoppers in these moments, Sephora created an app that allows them to scan an item in the store and immediately see product ratings and reviews on their phones. Having access to this information is that perfect new moment for customers to find everything they are looking for and get advice from Sephora.

Here are guidelines about how to create a moment-based program. *1. Map consumers' journey to understand their intent and context.* Ethnographic and observational studies are often more insightful for understanding the consumer journey than surveys or consumers' digital footprints, and for mapping the consumer journey, management typically focuses on the firm's product, not on the broader consumer journey. *2. Classify different moments into coherent groups.* Mapping the large number of micro-moments among consumers can seem daunting and impractical. Therefore, it is useful to categorize these moments into groups that are relevant for consumers and actionable for the firm. Google has classified micro-moments across four groups: *I want to know, I want to go, I want to do, and I want to buy.* Your business may have similar or different categories of moments, but it is useful to group hundreds of micro-moments into actionable and meaningful buckets. *3. Provide useful information.* Advertising is the art of providing valuable information when consumers need it. Technology enables us to identify these micro-moments of consumer need, and it is our task to provide useful information at those points in time. *4. Create stackable content.* Consumers have a very short attention span when they are looking for information on their smartphones. This situation calls for stackable content that addresses a specific intent of the consumer at that moment. *5. Speed matters.* These days consumers are always in a hurry and very impatient. If a website takes more than a few seconds to load, consumers get frustrated and leave the site. Based on an analysis of 900,000 mobile-ad landing pages across 126 countries, Google found that it takes 22 seconds to fully load a mobile landing page. And 53 percent of mobile-site visitors leave a page that takes longer than three seconds to load. Videos and images make it difficult for many sites to load faster, and often the cliché that less is more is applicable in these situations. Even though the tools of reaching consumers have changed in the digital era, we still need to have a deep understanding of consumers and provide value to engage them.

MAINTAIN LIFETIME CONSUMERS

Any business that wants to endure and profit must strive to reach the stage of lifetime consumers. Your organization must be unified in giving the consumer a great experience. All divisions and departments use a single view of the consumer covering all touch points. The consumer experience is aligned with organizational objectives. So, giving your consumer a great experience drives success in your organizational objectives. Here, most of your consumers are not only loyal to your organization, but they are also active advocates for you. The consumers think of themselves as customers for life. In fact, the leading digital businesses can take advantage of big and small data to reach this high level of consumer-relationship management.

The main objective of the lifetime consumer stays in three aspects. First is to create a great experience that retains consumers for life. Second is to optimize the consumer experiences across all online and offline channels, using real-time predictive automated personalization to offer the most relevant content. Third is to maintain competitive advantage by being the fastest and most agile in testing new initiatives.

Amazon is recognized as the world leader in digital excellence and e-commerce. Consumer experience is placed as the keystone of their vision. In order to make that happen they have data integration as a foundation, and then create consumer convenience through their use of predictive analytics. The American Customer Satisfaction Index rated Amazon as the leader in customer satisfaction with a rating of 88 in 2013. (The average rating for an Internet retailer is 78.) Focusing on continuous innovation and customer experience has not held Amazon back from financial growth. A key metric for online retailers is revenue per unique user. In 2011, estimates by JP Morgan found that Amazon generated $189 per unique user while the next closest competitors were eBay at $39, and Google at $24. That means Amazon generated more than seven times Google's revenue and was magnitudes ahead of other online businesses.

An organization like Amazon knows the consumer as an individual and has the ability to predict what the consumer wants. It makes the consumer-decision journey frictionless. Consumers find it easier to make decisions because predictive algorithms present the consumer's best or most probable decision paths in front of them. Consumers feel like they are being helped to make decisions rather than being sold. The organization appears to be a single entity with a single image and brand no matter whether the consumer approaches it offline, online, or through multiple channels. Although analysts used to say there is no loyalty on the Internet, that isn't true. Organizations with lifetime consumers have built a commitment and trust. Consumers are advocates for the organization and feel uncomfortable when forced to work with competitor brands. Major benefits come to an organization because lifetime consumers add significantly to the profitability by making repeat purchases, require minimal marketing cost, and advocate to bring in new consumers.

The leading organization with lifetime consumers can see an inflection point in customer growth. Marketing is easier because your consumers are advocating for you, and you can make accurate, data-driven decisions. Consumer growth is organic. A lot of growth comes from word of mouth that is not attributable to specific marketing programs. Your own customers are becoming advocates and spreading the word. Customers reach out to friends and associates to spread the word about what a great experience they have had in their relationship with your organization. They consider your organization as the first choice and accept alternatives only in special circumstances. Your consumers would feel a sense of loss if they had to deal with a different organization. At this point, your organization must keep its competitive advantage by maintaining a high speed of innovation, agility, testing, and redevelopment.

Marketing is data driven. It rapidly innovates, tests, and retries. Operational marketing uses analytics and testing for continuous improvement. Marketing software automatically makes recommendations on how to improve performance. Big data is used to identify unforeseen patterns, and consumers depend on your recommendations for their decisions. With accurate data, you can correctly predict growth and costs and have a fairly good estimate of Return on Marketing Investment for each channel and campaign. With the data, balancing cross-channel marketing and setting budgets is much easier. In a word, with lifetime consumers, the organization reaches the leading level that marketers aspire to.

CHAPTER 9
NEW ACTIONS

It's time for you to act in the real digital world. Go deeper, move from digital dabbling to digital strategy: Design the rulebook that's right for your business, choose your entry point, and develop your compelling narrative.

Now, you must lead.

TEST FAST, LEARN FAST, SCALE FAST

It should come as no surprise that leading digital giants are far more likely to invest in innovation: while 87 percent of digital giants said they invest in innovation, only 38 percent of traditional incumbents do so. Being innovative is not simply about doing innovative things. Instead, it is about cultivating an organizational environment that is conducive to innovation. It is about being open to new ideas, wherever they may be found. Leading digital companies were more than twice as likely as traditional companies to encourage new ideas to be shared and tested at all levels of the organization. More important, being innovative is also about a willingness to act on those ideas. Innovation is critical to business success in a rapidly changing environment. In practice, most organizations that were not leading in the digital age struggle mightily with innovation for two reasons. First, the culture of most traditional organizations has evolved to reduce or eliminate the variations necessary for the experimentation that leads to innovation. And second, leaders find it challenging to innovate while also keeping the company's core businesses running efficiently and effectively.

The biggest challenge affecting an organization's ability to compete more effectively in a digital environment is experimentation and getting people to take risks. This barrier is the highest even for leading digital companies. We have seen traditional companies struggle with the need for experimentation because they are driven by a fear of failure. The efforts at variance reduction are an effective strategy in traditional incumbents, where the organizational goals and conditions are stable. Yet companies can only achieve this level of precision by optimizing to a certain set of conditions that are well understood. That is the type of approach to eliminating variances used in the Six Sigma method. In more turbulent environments—where the conditions under which business takes place are rapidly changing—efforts at variance reduction may not be ideal. They may have the benefit of identifying new business processes or opportunities that are far more effective or efficient, but a possible unintended consequence is that the organization may not be able to realize these benefits because it is optimized around the old conditions. In many ways, a culture of experimentation is about intentionally creating variance in existing processes to see if there are better ways of doing things. Certainly, Google is quite different from most companies. Its purpose is to literally change the world. Their offering lends themselves to this type of experimentation, and they have abundant resources to devote to these efforts.

Test fast, test small, test enough. A primary reason why many organizations struggle with experimentation is that they have conditioned themselves to believe that failure is anathema. No wonder most organizational employees are nervous about experimenting if the company's tolerance for failure is the Six Sigma threshold (failing 3.4 times out of every 1 million attempts). You cannot hope to successfully experiment at that rate. In the digital age, how companies deal with setbacks may determine their ability to survive. Because new challenges are becoming the norm, and so much is unknown and untested, failure is inevitable. So, a critical factor in helping organizations become better experimenters is helping them become better at testing ideas, learning from these tests, and scaling quickly when the tests reveal productive insights. It may be easier to focus the organization on adopting a test and learn mindset, rather than a fail-fast one.

A good way to test fast is to put a fixed short-term timeline on experiments. Experimental organizations often perform short springs in which they attempt to change one aspect of the organization. At the end of the spring, the experiment concludes, and the success or failure of the experiment is determined. This established time frame provides managers an easy decision point to stop or refocus the experiment, instead of allowing stumbling projects to be drawn out for long periods.

An important afterthought of testing fast is to test small. A company wouldn't want a multibillion-dollar IT implementation project to fail just to learn some important lessons. So, the company must ensure that it creates acceptable parameters for experimentation and learning to take place. Be sure you set up small experiments that limit the damage from failures, allowing you to learn and move on.

Last is testing enough. Companies need to manage risk as a portfolio and keep failure within a certain tolerance level. Is the right failure rate 10 percent or 90 percent? Make sure to find your organization's Goldilocks Zone for testing. Yet, managers should also remember that if they aren't failing enough, they might not be bold enough. For example, one division of the U.S. Department of Agriculture has an established risk-tolerance threshold for digital projects. If they aren't failing enough, they increase the ambitiousness of the projects to fail more often. Not many people think of the government as a place to look for innovative practices, but we think that many companies can benefit from an environment of testing enough.

Learn faster. Though the word failure has a negative implication, conversation around it is starting to shift. Even so, all the talk about organizations needing to fail fast emphasizes the speed aspect and deemphasizes the learning aspect. The idea is not simply to fail quickly and move to the next idea. Insights must be gathered from the failure to make it valuable. Just knowing that "A didn't work, so let's try B" is not enough. Understanding why A didn't work is where the insights are found and where the learning happens for the organization. As a scientist will tell you, they conduct experiments to test hypotheses and ideas. By proving or disproving their hypotheses, they gain knowledge. In a similar way, organizations should approach experimentation with a goal to learn. From this perspective, testing becomes valuable as input, or insight, into what didn't work and what could have been done better. The key is to not get stuck in these setbacks and instead learn from them and move on. The implication of this approach is that you should also learn from projects that succeed. If you don't understand why a particular experiment succeeds, then you don't know whether—and how—to export lessons from the success, which may be just as important as the success itself. Success is not the short-term goal, but learning is. Just as learning is essential to a digital talent mindset, it should also be a key part of the organizational mindset.

The process of learning at the organizational level can take many different forms. For some, it can involve formal after-action reports or sessions that debrief the results of an experiment to extract lessons learned. For others, it may involve all hands meetings where project teams present ongoing results and gain feedback from members of other teams. It may also involve splitting the team up and incorporating its members into new teams to combine employee knowledge in new ways. In short, organizations can learn from their successes and failures in several different ways; the key is to clearly identify and intentionally engage in processes that help the organization learn from its experience.

Iterate, Iterate, Iterate. As scientists remind us, lessons learned should not occur in a vacuum. Instead, learning during one round of experimentation should inform the types of experiments companies engage in during the next round of innovation. The experiments and failures should not be scattershot, unrelated projects, but targeted toward a specific overarching goal. This discipline becomes even more important in the digital age because of the opportunity for machine learning and AI. There is great value in having some kind of rich reproducibility tracking platform that can trace nearly every aspect of each interaction of new models to ensure that results can be deployed accurately and efficiently. Just as innovation is more about the environment of cultivation ideas at all levels of the organization, so too, the spirit of feedback and learning should pervade all levels. Here, the difference between leading digital companies and others is even more extreme. Iteration involves taking the lessons learned from the previous experimentation effort and building that knowledge into the next set of experiments. The lessons from experiments may also be at a meta-level, informing the experimentation process.

Scale fast. Another danger with the fail-fast approach is that companies are satisfied with simply running experiments and think this is enough to embrace risk. Yet digital initiatives aren't risky if they never promise to touch your core business. Moreover, considerable research suggests that innovation has little real effect on business unless companies fundamentally transform their business models. Experimentation just for the sake of experimentation isn't particularly valuable. Indeed, the key differentiator of innovation at leading digital companies, and others, may not be their ability to experiment well and learn from their experiments. The more telling differentiator may be their willingness to take the lessons learned from both successful and failed experiments and scale those experiments across the organization to drive business model transformation.

The reason that companies struggle with innovation is that they can't afford to drop everything and shift their focus to learning and experimenting with new technology. They must innovate in a way that allows their core business to function while also influencing that core business. It equates with trying to fix a plane in midflight. It is extremely hard to be a digital leader in the current environment when you are being asked to change the wings on a plane while it's flying, with total security, reliability and stability, as well as complete innovation, at the same time. Leading digital companies have found ways to be innovative amid the need to maintain business operations. As companies begin to experiment more, this increase puts additional pressure on the need to balance experimentation with effective exploitation. Organizations need to find new ways of doing business through exploration and experimentation while also maintaining a viable business and exploiting established competencies.

GREATER STAKES, GREATER THREATS, GREATER REWARDS

Consider these facts: The average lifespan of an S&P 500 corporation in the 1950s was 60 years, by 2012 it was less than 20 years. Half of the companies that composed the Fortune 500 in 2000 are no longer on the list today. From 2013 to 2017, median CEO tenure at large corporations dropped by one full year, from six to five years. Acquisitions, mergers, privatization, and bankruptcies have decimated the status quo. These trends will accelerate. The technologies driving digital transformation are now robust, mature, and widely available. Digital leaders will emerge across industries, harnessing AI and IOT to achieve step-function improvements in their business processes and outcompete slower rivals.

At the same time, capital markets become more efficient every day, adding to the pressure on underperforming companies which will be treated ruthlessly. Hawk-eyed hedge fund and private-equity professionals constantly search for opportunities to acquire, merge, break up, and liquidate corporations that show traces of vulnerability. The private-equity industry manages $2.5 trillion, with nearly $900 billion in dry powder waiting to be deployed. This capital will not only be used in takeover bids. It will also be used to fund rapidly growing digital-native competitors. Competitive threats on the digital landscape can come from any and every direction, in unexpected ways. In January 2018, Amazon and JP Morgan Chase announced a joint venture into the health-care industry. In a single day of trading, $30 billion of market capitalization was erased from the 10 largest US health care companies, with some dropping by as much as eight percent. Although the threat of extinction is great and the need for action urgent, companies that succeed will be richly rewarded. Every significant study of the potential economic impact of digital transformation shows that organizations have the opportunity to create trillions of dollars of value through AI and IOT. Companies that transform will be operating on an entirely different level from their lagging competitors. Digital transformation is the next do-or-die imperative. How company leaders respond will determine whether their companies thrive or perish.

The transformative leaders. As we've noted previously, digital transformations are both pervasive and go to the core of the companies, the capabilities and assets that define and differentiate the business. For those reasons, all senior leaders and business functions must drive digital transformation from the top down. In fact, digital transformation has completely inverted the technology adoption cycle that prevailed in prior decades. Previously, new technologies frequently came out of research labs; new companies were formed to commercialize the technologies, and over time they were introduced to industry through the IT organization. And eventually, after gradual adoption, they gained the attention of the CIO. Only after years did they reach the CEO's desk, and this was usually only in the form of a brief or a budget approval requiring the CEO's signature. Today, CEO's initiate and mandate digital transformations. This is a big change. It is an entirely new paradigm for innovative technology adoption, driven by both the existential nature of the risk, as well as the magnitude of the challenge. Whereas the CEO was generally not involved in IT decisions and strategy, today he or she is the driving force.

Rather than a sequential, step-by-step process, there are principles to guide the transformation. Some actions will be taken in parallel; some in sequence, and every organization will adapt to fit their particular situation. All of the points are essential, as they touch on the key areas of leadership, strategy, implementation, technology, change management, and culture.

Marshal the chief officer team as the digital transformation engine. A leadership team committed to the digital transformation agenda is an absolute requirement and a first priority. C-suite must become the engine of digital transformation. Don't take this to mean the CEO or CMO are suddenly writing code or wrangling new technologies. It does mean that the CEOs and other top-level executives know what opportunities digital opens up for their company and how to create a digital value proposition that distinguishes their business from others. This is a change from previous eras when CEOs required only a rudimentary understanding of how technology worked. Today, CEOs have to keep up with a deluge of information about ever-changing technologies, able to decide what is relevant to the business and prioritize which new technologies to focus on, and which to filter out. The senior chief experience offer (CXO) team needs to marshal the funding, resources, and relationships necessary to enable digital transformation. Reinventing a company requires this commitment from the C-suite to ensure the entire workforce is aligned behind the vision. The mission to own and drive digital transformation, as a core evolution across the business, was clear: transform the company for agility, insight, and growth. But each business unit had individual priorities and unclear understanding initially of what digital transformation meant. Only after establishing a common understanding of the digital transformation mandate, and clarity about its goals, was the CEO successful in getting his CXO on board. Digital transformation requires adopting a long-term perspective. It requires moving beyond just measuring financial performance for the next quarter to also thinking toward a broader, bigger picture of the future and how the company will fit within it. It also requires a certain personality type. Leaders need to be able to handle risk, they need to be willing to speak out, and they need an experimentation mindset. Leaders must also be comfortable with technology and conversant with technological terms and concepts. This means spending time to understand the capabilities of relevant technologies and what the development teams are doing. CEOs need to surround themselves with a C-suite and a board that share these traits in order to propel transformation.

Appoint a CDO with authority and budget. While the entire C-suite must be the driving force behind the digital transformation agenda, their needs to be a dedicated senior executive singularly focused on digital transformation results. A chief digital officer (CDO) is empowered with authority and budget to make things happen. The CDO's primary role is chief evangelist and enabler of digital transformation, the one who focuses on the transformation strategy and who communicates across the organization on the action plans and results. The CDO's role is also to think about what's next and how the organization needs to evolve in order to seize new opportunities, create new value for customers and the business, and manage potential risks and disruptions. The CDO role is an important one, but insufficient to broker all the functional innovation that needs to happen across the organization in order to transform. Best practice also requires a central organization to act as the hub of digital transformation, which is a cross-functional team of software engineers, data scientists, product specialists, and product managers who work collaboratively in an enterprise to develop and deploy AI and IOT applications. The CEO and CDO play key roles in forming, supporting, and engaging with the hub. The CDO may recommend complementing internal capabilities with outside partner help. This can be a useful strategy to jump-start a digital transformation initiative. The CDO needs to have the full support of the CEO and the clear authority to assume responsibility for the digital transformation road map, vendor engagement, and project supervision. The CDO should act as the CEO's full-time partner, responsible and accountable for the result.

Get work done and capture business value. Just as vital as assembling and aligning internal forces is the need to capture business value as soon as possible. Many organizations get complex data-lake projects for years at great expense and yield little or no value. Trying to solve the issue of building an all-encompassing data lake for analytics and insight is an unfortunately common occurrence. GE spent over $7 billion trying to develop its digital transformation software platform, an effort that ultimately contributed substantially to the failure of the company and the replacement of its iconic CEO. The corporate landscape is littered with lengthy, expensive IT experiments that attempted to build digital capabilities internally. Perhaps you should leave that to experts with demonstrated track records of delivering measurable results and ROI. The way to capture business value is to work out the use case first, identify the economic benefit, and worry about the IT later. While this may sound like heresy to a CIO, a use-case-first model allows for focus on the value drivers. By adopting a phased delivery model, this is essentially the agile development model popular in software development today in which teams can achieve faster results. With a phased model, projects get addressed and delivered in short, iterative cycles aimed at continuous incremental improvement, each contributing additional economic benefit, enabling teams to focus on the end result and the customer. For team members, this has the psychological benefit of helping them feel involved in productive efforts that contribute to company growth, instilling motivation around concrete digital transformation efforts.

Forge a strategic vision in parallel and go forward. Digital transformation strategy should be focused on creating and capturing economic value. A proven approach is to map out your industry's full value chain, and then identify the steps of this value chain that have been, or that you expect to become, digitalized. This will help you understand where your gaps are. Your value-chain map and your strategy might initially center on inventory optimization, production optimization, AI predictive maintenance, and customer churn. How you sequence your strategy depends on how and where you can find business value. You should sequence these projects in the order that offers the highest probability of delivering immediate and ongoing economic and social benefit. Tackle these projects in a phased approach proving out your strategy, fine-tuning your processes, and adding value to the company incrementally with each project deliverable. Two key elements of developing your strategy are benchmarking and assessing the forces of disruption in your industry. Benchmarking, like all aspects of business digital transformation, happens in a competitive context. You will want to benchmark your digital capabilities against those of your peers and use best practices. Where is your industry as a whole on the digital maturity spectrum? Who are the digital frontrunners in your industry? How do you compare with them? These are the critical questions you will need to answer to get an effective lay of the land. Once you've taken stock of your organization's digital capabilities, you can begin to try to understand how your industry might be changing and how you need to be prepared for that future. Clearly identify the existential threats to your company in the coming decade. Think about alternatives to turn those threats into areas of strategic competitive advantage. The other is assessing the forces of disruption. Developing your strategy requires an assessment of your industry and the forces of disruption likely to shake it up. It means identifying threats not only from known competitors, but also from unexpected areas. These could be competitors pursuing a higher-quality approach, low-cost upstarts, more nimble digital natives, companies that provide more visibility, or existing entities expanding into new areas. Understanding where your company and your industry sit in terms of their susceptibility to disruption will help guide critical strategic choices. The right time to start taking control of your unique state of disruption is now.

Draft a digital transformation roadmap and communicate it to stakeholders. It's time to involve building a portfolio of opportunities, identifying and prioritizing functions or units that can benefit most from transformation. It also involves locating and starting to address roadblocks to transformation. During the design phase, companies also invest in framing and communicating the vision for the transformation to build support for needed changes, and they invest in systems to industrialize data analytics, making a resource for every operation. First, clearly define a future vision for your digital business. What does your ideal future state look like in terms of your organizational structure, people and leadership, product and services, culture, and adoption of technology? Use this ideal future state to compare against your current state and zoom in on any gaps. Put your transformation on a timeline with clear milestones. The best digital transformation roadmap involves concrete plans and timeliness to bring advanced AI applications into production. The development and rollout pace will depend on your specific objectives and circumstances. Organization alignment is critical. As a business leader, you need to communicate effectively and sell your vision to stakeholders across the organization. Changing the organization, its culture, and its mindset, requires buy-in across production lines and stakeholders, not just at the C-suite level.

Choose your partners carefully. To fulfill your vision of digital transformation, it's vital to pick the right partner. This applies to the full ecosystem of partnerships; the CDO and CEO need to establish software partners, cloud partners, and a spectrum of other partnerships and alliances. In a digital transformation world, partners play a bigger role than in the past. Three key areas where partners add significant value: strategy, technology, and services. Strategy means consulting partners can help flesh out your strategy: map your value chain, uncover strategic opportunities and threats, and identify key applications and services you will need to develop in order to unlock economic value. Technology means software partners provide the right technology stack to power your digital transformation. As you scale your transformation, the complexity of your system will grow exponentially. Each individual application or service will require cobbling together a large number of low-level components, and your engineers will spend a majority of their time working on low-level technical code rather than solving business problems. It is good to seek out technology providers that can offer a cohesive set of higher-level services for building advanced applications with large data volumes. Services means to look for professional services partners who can help build your advanced application and augment your staff. They can provide teams of developers, data integration specialists, and data scientists if you do not have those in-house talent profiles.

Focus on economic benefit. As you approach the digital transformation you should stay focused on economic and social benefits, meaning the benefits for your customers, for your shareholders, and for society at large. If you and your team cannot identify projects for digital transformation that return significant economic value within one year, keep looking. If the solution cannot be delivered within a year, don't do it. This is a team sport. Engage with your management team, your business leaders, and your trusted partners. If you make the effort to wrestle with enough ideas, one project will emerge from the process that is tractable, offers clear and substantial economic benefit, can be completed in six months to a year, and once completed successfully, presents the opportunity to be deployed and scaled across the company. When you have identified one or more such projects, bring in experts to provide the software technology, and who fully support the project. Review progress weekly. Set clear and objectively quantifiable milestones, and mandate that they be met. When a milestone slips, require a mitigation plan to get it back on schedule. Walk around daily to talk to the people doing the work so you can feel the progress. And most importantly, assume that unless you do all of the above, the project will fail.

Create a transformative culture of innovation. A CEO may have a clear vision of what needs to happen to transform the organization. But senior management, middle management, and rank-and-file employees must also fully understand that vision and operate in an environment conducive to success. In order to effectively drive digital transformation, CEOs need to understand what the world of digital transformation and disruption looks like, and what digital products can do. As part of this effort to gather information and inform their own visualizations of their company's future, CEOs should conduct corporate executive visits at the sources of disruption. Think of Uber, Airbnb, Amazon, Apple, Tesla, and Netflix. These companies outcompete traditional players, disrupt entire industries and create new business models. They've created entirely new ways of approaching their business and services. From these companies, corporate leaders can learn what it means to have an innovation culture. This entails cultivating a culture with a core value, one that rewards collaboration, hard work, and continuous learning. The way you attract, the way you retain, and most importantly, the way you motivate and organize people today has changed dramatically from what we experienced in recent decades. With the baby boomers, we could focus primarily on compensation plans to motivate individuals. Today, we oversee multi-generational workforces comprised of a complex tapestry of baby boomers, Gen Xers, and millennials, very rich and complex groups. Individuals in today's workforce have divergent value systems and diverse motivations and skill sets. You need to figure out how to take that powerful mix of diverse attitudes, goals, and motivations and turn it into something cohesive, focused, and productive, with a shared mission and common purpose.

The companies able to innovate effectively are those that share certain characteristics: high tolerance for risk, agile project management, empowered and trained employees, collaborative culture, and an effective decision-making structure. Without a culture that encourages innovation and risk-taking, even the best, thought-out digital transformation strategy will fail.

CONCLUSION

DIGITAL TRANSFORMATION IS NOT ABOUT UPDATING YOUR TECHNOLOGY, IT'S ABOUT UPDATING YOUR STRATEGIC THINKING, BEING FULLY IN CHARGE OF YOUR CALL TO ACTION.

It is important that digital transformation is not an end in and of itself. With the rapid evolution of digital technology, there will be an ongoing process of education, experimentation, and delivery. Success in transforming organizations demands everyone in the organization is digitally aware in the same way, and within the diversity of their roles in driving corporate strategy.

This book has given you a practical understanding of digital transformation powered by cloud computing, big data, AI and IOT. As in evolutionary biology, the new environment created by the confluence of these technologies poses an existential threat for companies from the traditional industrial era, but also creates a massive opportunity for those that take advantage of innovative resources. Companies that recognize the magnitude of the opportunity, and are willing and able to adapt, will be well positioned to unlock substantial economic value.

The next two decades will bring more information technology innovation than the past half century. New business models will emerge. Products and services unimaginable today will be ubiquitous. New opportunities will shape the long-term vision and daily operation of companies. Getting started is often the hardest part, but you have to be ready to get through it. Be brave. Embrace the brighter future of the new digital world.

Notes

INTRODUCTION

1. Nicholas Negroponte, *Being Digital*, Alfred A. Knopf Inc., New York, 1995.
2. Alexander Osterwalder, Yves Pigneur, *Business Model Generation*, Wiley & sons, New Jersey, 2010.
3. Gerald C. Kane, Anh Nguyen Phillips, Jonathan R. Copulsky, Garth R. Andrus, *The Technology Fallacy: How People Are The Real Key to Digital Transformation*, The MIT Press, London, 2019.
4. Sunil Gupta, *Driving Digital Strategy: A Guide to Reimagining Your Business*, Harvard Business Review Press, Boston, 2018.
5. Venkat Venkatraman, *The Digital Matrix: New Rules for Business Transformation Through Technology*, LifeTree Media, Vancouver, 2017.
6. Thomas M. Siebel, *Digital Transformation: Survive and Thrive in an Era of Mass Extinction, Rosetta Books*, New York, 2019.
7. Michael Scott Morton, *The Corporation of the 1990s: Information Technology and Organization Transformation*, Oxford University Press, New York, 1991.
8. David Roger, *The Digital Transformation Playbook,* Columbia Business School Publishing, New York, 2016.
9. George Westerman, Didier Bonenet, Andrew McAfee, *Leading Digital: Turning Technology into Business Transformation,* Harvard Business School Press, Boston, 2014.
10. Laure Claire Reillier, Benoit Reillier, *Platform Strategy: How to Unlock the Power of Communities and Networks to Grow Your Business*, Routledge, New York, 2017.

11. Pedro Domingos, *The Master Algorithm: How the Quest For the Ultimate Learning Machine Will Remark Our World*, Levine Greenberg Rostan Literary, 2015.

12. Mark J. Perry, *Fortune 500 Firms in 1955 v 2015*, American Enterprise Institute blog, 2016

13. Interbrand, *Best Global Brand 2015*, Interbrand.com 2016.

14. William Edwards Deming, *The New Economics: For Industry, Government, Education*, MIT Press, Cambridge, 2000.

15. Everett M. Rogers, *Diffusion of Innovations*, Free Press, New York, 1962.

16. Carol Hymowitz, *Innovation, Wall Street Journal*, 2006.

CHAPTER 1

1. Alfred D. Chandler, *Scale and Scope: The Dynamics of Industrial Capitalism*, Belknap Press, Cambridge, 1990.

2. William Bruce Cameron, *Information Sociology: A Casual Introduction to Sociological Thinking*, Random House, New York, 1963.

3. Steve Blank, *Why the Lean Start-up Changes Everything*, Harvard Business Review, 2013.

4. Danny Sullivan, "Google Annual Search Statistics: 100 Billion Searches Per Month", *Statisticbrain.com*, 2016.

5. Simon Van Zuylen, *The Struggles of New York City Taxi King*, Bloomberg Business Week, 2015

6. Mark Zuckerberg, "The Hacker Way", *WIRED*, 2012.

7. Brian Olsavsky, *Amazon Management on Q4 2015 Result: Earnings Call Transcript*, Seeking Alpha, 2016.

8. Netflix Tech Blog, "It's All About Testing: The Netflix Experimentation Platform", *Techblog.Netflix.com*, 2016.

9. Mike Isaac, Michael Merced, "Uber Turns to Saudi Arabia for $3.5 Billion Cash Infusion", *The New York Times*, 2016.

10. Ray Kurzweil, "The Law of Accelerating Returns", *Kurzweilai.net*, 2001.

11. Dale Evans, *The Internet of Things: How the Next Evolution of the Internet Is Changing Everything*, IBAC, San Jose, 2011.

12. Thomas Hout, George Stalk, *Time-Based Results*, Boston Consulting Group, 2016.

13. James Manyika, *Harnessing Automation for a Future That Works,* McKinsey Global Institute, 2017.

14. Jeffrey Pfeffer, Robert Sutton, *The Knowing Doing Gap,* Harvard Business School, Boston, 1999.

CHAPTER 2

1. Geoffrey G. Parker, Marshall W. Van Alstyne, *Platform Revolution: How Networked Markets are Transforming the Economy and How to Make them Work for You*, W.W. Norton & Company, New York, 2016.

2. Frederick W. Taylor, *The Principles of Scientific Management*, Harper & Brothers, New York, 1911.

3. Michael E. Porter, James E. Heppelmann, *How Smart, Connected Products are Transforming Competition*, Harvard Business Review, Boston, 2014.

4. Venkat Venkatraman, *What Comes After Smart Products*, Harvard Business Review, Boston, 2015.

5. Robert Tercek, *Vaporized: Solid Strategies for Success in a Dematerialized World*, LifeTree Media, Vancouver, 2015.

6. Thought Machine, *Introducing Vaultos: New Banking Model,* Beingtechnologies.com, 2016.

7. McKinney, *GE's Jeff In melt on Digitizing in the Industry Space,* McKinsey & Company, 2015.

8. Jack Nicas, Jeff Bennett, "Alphabet, Fiat Chrysler in Self-Driving Cars Deal", *The Wall Street Journal*, 2016.

9. Ingrid Lunden, *VW Invests $300M in Uber Rival Gett in New Ride-sharing Partnership*, Tech Crunch, 2016.

10. Sunil Gupta, Sara Simonds, *Goldman Sachs' Digital Journey*, Harvard Business School, Boston, 2017.

11. Michael E. Porter, James E. Heppelmann, *How Smart, Connected Products Are Transforming Competition*, Harvard Business Review, 2014.

CHAPTER 3

1. Warren Berger, *How Brainstorming Questions, Not Ideas, Sparks Creativity*, Fast Company & Inc, 2016.
2. N. Kulatilaka, N. Venkatraman, "Strategic options in the Digital Era", *Business Strategy Review*, 2001
3. Chris Anderson, "The End of Theory: The Data Deluge Makes the Scientific Method Obsoletes", *WIRED*, 2008
4. Xavier Amatriain, Justin Basilico, "Netflix Recommendations: Beyond the 5 Stars", Netflix Tech Blog, 2012
5. Mark McClusky, "The Nike Experiment: How the Shoe Giant Unleashed the Power of Personal Metrics", *WIRED*, 2009
6. Brian Dolan, *10 Reasons Why Google Health Failed*, MobiHealth News, 2011
7. Arielle Duhaime, "Apple Dives Straight into Health Care with Release of First CareKit Apps", theverge.com. 2016.
8. Yongdong Wang, *Your Next New Friend Might Be a Robot*, Nautilus, 2016.
9. Dsamaniego, *Step Inside the Eureka Innovation Lab*, Levi Strauss & CO, 2014.
10. Cawood, *Streamlining*, Farm Weekly, 2015.
11. Sarwant Singh, "The Future of Car Retailing", *Forbes*, 2016.
12. Tanguy Catlin, *A Roadmap for a Digital Transformation*, McKinsey, 2017.
13. Brian Solis, *The Six Stages of Digital Transformation Maturity*, Altimeter Group and Cognizant, 2016.

CHAPTER 4

1. Peter James, Arnoud Meyer, *Ecosystem Advantage: How to Successfully Harness the Power of Partners,* California Management Review, 2012.
2. Erle Ellis, *Ecosystem: The Encyclopedia of the Earth,* Eoeearth.org.com, 2014.

3. Felix Oberholzer Gee, Julie Wulf, *Alibaba's Taobao,* Harvard Business School, Boston, 2009.

4. Geoffrey Parker, Marshall Alstyne, *Two-sided Network Effects: A theory of Information Product Design,* Management Science, 2005.

5. Elizabeth Altman, Mary Tripsas, *Product-to-Platform Transitions: Organizational Identity Implications,* Oxford University Press, Oxford, 2015.

6. Chan Kim, Renee Mauborgne, *"What Is Blue Ocean Strategy?",* Blue Ocean Strategy.com, 2016.

7. Geoffrey A. Moore, *Escape Velocity: Free Your Company's Future from the Pull of the Past,* Harper Business, New York, 2011.

8. William Boston, Eric Sylvers, "Auto Makers Gear up to Take on the Challenge from Google and Apple", *The Wall Street Journal,* 2015.

9. Lulu Yilun Chen, "Apple's Cook Struck $1 Billion Deal with China's Didi in 22 days", *Bloomberg,* 2016.

10. Stephen Pulvirent, *TAG Heuer, Google Release First Swiss Luxury Smart watch: All the Details,* Bloomberg Pursuits, 2015.

11. Adam Bryant, "Chief of Microsoft, On His New Roles", *The New York Times,* 2014.

12. David S Evans, Richard Schmalensee, *Matchmakers: The New Economics of Multisided Platforms,* Harvard Business Review Press, Cambridge, 2016.

13. Thomas Eisenmann, *Platform-Mediated Networks: Definitions and Core Concepts,* Harvard Business School, Boston, 2006.

CHAPTER 5

1. Adam Brandenburger, Barry Nalebuff, *Co-Opetition: A Revolutionary Mindset that Combines Competition and Cooperation,* Currency Doubleday, New York, 1996.

2. Theodore Levitt, *Marketing Myopia,* Harvard Business Review, 1960.

3. Germany Trade and Invest, *Industries 4.0: Smart Manufacturing for the Future,* Germany Trade and Invest, Berlin, 2014.

4. Cornelius Baur, Dominik Wee, *Manufacturing's Next Act*, McKinsey & Company, 2015.

5. Timothy F. Bresnahan, Shane Greenstein, *Technological Competition and the Structure of the Computer Industry*, Journal of Industrial Economics, 1999.

6. Gautham Nagesh, "Mary Barra's Road Map for GM Centers on Customer Data, Connectivity", *The Wall Street Journal*, 2015.

7. John C. Camillus, *Wicked Strategies*, University of Toronto Press, Toronto, 2016.

8. Horst W. Rittel, Melvin M. Webber, *Dilemmas in a General Theory of Planning*, Policy Science, 1973.

9. Ranjay Gulati, *Reorganize for Resilience: Putting Customers at the Center of Your Business*, Harvard Business Review Press, Cambridge, 2010.

10. Steve Lohr, *Data-Ism: Inside the Big Data Revolution*, Harper Business, New York, 2015.

11. Tom Warren, *Lenovo Refused to Sell Microsoft's Surface Because They're Competitors*, The Verge, 2015.

12. Paul A. Pavlou, Omar A. Sawy, *The Third Hand: IT-Enabled Competitive Advantage in Turbulence through Improvisational Capabilities*, Information Systems Research, 2008.

CHAPTER 6

1. Carmine Gallo, Steve Jobs: *The World's Greatest Storyteller*, Forbes, 2015.

2. Joseph A. Nye, *Soft Power: The Means to Success in World Politics*, Public Affairs, New York, 2004.

3. Virginia M. Rometty, *Chairman's Letter*, IBM Annual Report, 2016.

4. Sarya Nadella, *The Partnership of the Future*, Slate, 2016.

5. Carl B Frey, Michael A. Osborne, *The Future of Employment: How Susceptible are Jobs to Computerization*, Oxford Martin School Working Paper, 2013.

6. Tom Davenport, Julia Kirby, *Only Humans Need Apply: Winners & Losers in the Age of Smart Machines*, Harper Business, New York, 2016.

7. Bureau of Labor Statistics, US Department of Labor, Economic news release, 2016.

8. Jerry Kaplan, *Humans Need Not Apply: A guide to Wealth and Work in the Age of Artificial Intelligence*, Yale University Press, New Haven, 2015.

9. Gary Kasparow, *The Chess Master and the Computer*, New York Review of Books, 2010.

10. Erik Brynjolfsson, Andrew McAfee, *The Second Machine Age: Work Progress and Prosperity in a Time of Brilliant Technologies*, W.W. Norton & Company, New York, 2014.

11. Mike Ramsey, Toyota Hires Entire Staff of Autonomous Vehicle Firm, *The Wall Street Journal*, 2016.

12. David Nadler, Michael Tushman, *A Model for Diagnosing Organizational Behavior*, Organizational Dynamic, 1980.

13. Laszlo Bock, *Work Rules Insights from Inside Google That Will Transform How You Live and Lead*, Hachette Group, New York, 2015.

14. Kevin J. Delaney, *No One Should Have The Word Strategy in Their Job Title*, Quartz, 2016.

CHAPTER 7

1. Paul Leonardi, *When Does Technology use Enable Network Change in Organization? A Comparative Study of Feature Use and Shared Affordance*, MIS Quarter, 2013.

2. Peter F. Drucker, *The Theory of the Business*, Harvard Business Review, 1994.

3. James G. March, *Exploration and Exploitation in Organizational Learning*, Organization Science, 1991.

4. Eamonn Kelly, *Introduction: Business Ecosystems Come of Age*, Deloitte Insights, 2015.

5. Ansoff Declerck, R. Hayes, *From Strategic Planning to Strategic Management*, John Wiley & Sons, London, 1976.

6. Sunil Gupta, *Big Data: Big Deal or Big Hype*, The European Business Review, 2015.

7. Glen Urban, *Morphing Banner Advertising,* Marketing Science, 2013.

8. Louis M. Gerstner, *Who Says Elephants Can't Dance:* Inside IBM's Historic Turnaround, Harper Collins, New York, 2002.

9. Clayton M. Christensen, *The Innovator's Dilemma: When New Technologies Cause Great Firms to Fail,* Harvard Business Press, Boston, 1997.

10. Lawrence M. Fisher, "Preaching Love Thy Competitor", *The New York Times,* 1992

11. Walter Isaacson, *Steve Jobs,* Simon & Schuster, New York,   2011.

12. Charles Bruno, *Big Red Keeps Rolling,* Network World, 1995.

13. Sir Martin Sorrell, *What We Think,* WPP Annual Report, 2012.

14. Eric Von Hippel, *The Dominant Role of Users in the Scientific Instrument Innovation Process,* Research Policy, 1976.

15. Eric Von Hippel, *Free Innovation,* MIT Press, Cambridge, 2017.

16. Dietmar Harhoff, Karim Lakhani, *Revolutionizing Innovation: User, Communities, and Open Innovation,* MIT Press, Cambridge, 2016.

17. James Gibson, *The Ecological Approach to Visual Perception,* Houghton Mifflin Harcourt, 1979.

18. Uzi Shmilovici, *The Complete Guide to Freemium Business Models,* Techcrunch, 2011.

CHAPTER 8

1. David Court, *The Consumer Decision Journey,* McKinsey Quarterly, 2009.

2. Sunil Gupta, Joseph Davies Gavin, *BBVA Compass: Marketing Resource Allocation,* Harvard Business School, Boston, 2012.

3. Corrie Driebusch, Eliot Brown, "Blue Apron Serves Up an Insipid Offering", *The Wall Street Journal,* 2017.

4. Gregory Karp, *Millennial, Nontraditional Travelers Chase the Sapphire Reserve,* Nerd Wallet, 2017.

5. Carl Mela, Sunil Gupta, Donald Lehmann, *The Long-term Impact of Promotion and Advertising on Consumer Brand Choice,* Journal of Marketing Review, 1997.

6. Justina Perro, *Mobile Apps: What's a Good Retention Rate?* Localyics.com, 2017.

7. Thales Teixeira, Morgan Brown, Airbnb, Etsy, *Uber: Acquiring the First Thousand Customers,* Harvard Business School, Boston, 2016.

8. Philipp Schmitt, Bernd Skiera, *Why Customer Referrals Can Drive Stunning Profit,* Harvard Business Review, 2011.

9. Arun Sundararajan, *The Sharing Economy: The End of Employment and the Rise of Crowd-based Capitalism*, The MIT Press, Cambridge, 2016.

10. Bala Iyer, ChiHyon Lee, N. Venkatraman, *Managing in a Small World Ecosystem: Lessons from the Software Sector,* California Management Review, 2006.

11. A Guess, *How PayPal Is Using Deep Learning to Root Out Fraud,* Dataverisy, 2015.

12. Andrew Grove, *Only the Paranoid Survive: How to Exploit the Crisis Points That Challenge Every Company,* Doubleday Business, New York, 1996.

13. Jeroen P. Jong, *Market Failure in the Diffusion of Consumer Developed Innovations: Patterns in Finland,* Research Policy, 2015.

14. Dietmar Harfoff, "Context, Capabilities, and Incentives: The Core and the Periphery of User Innovation," in *Revolutionizing Innovation*, MIT Press, 2016.

15. Allison Mooney, Brad Johnsmeyer, *I Want-to-Buy Moments: How Mobile Has Reshaped the Purchase Journey,* Google, 2015.

CHAPTER 9

1. Mark Roberti, "How Tiny Wireless Tech Makes Workers More Productive", *The Wall Street Journal*, 2016.

2. Mathieu Dougados, *The Missing Link: Supply Chain and Digital Maturity,* Capemini Consulting, 2013.

3. Claude Bernard, An Introduction to the Study of Experimental Medicine, Henry Schuman Inc, 1949.

4. Howard Yu, *What Pokemon Go's Success Means for the Future of Augmented Reality,* Fortune, 2016.

5. Manu S. Mannoor, Ziwen Jiang, *3D Printed Bionic Ears*, Nano Letter, 2013.

6. Gordon R. Sullivan, Michael V. Hopper, *Hope is Not a Method: What Business Larders Can Learn from America's Army*, Broadway Books, New York, 1997.

7. Duncan Watt, Jonah Peretti, *Viral Marketing for the Real World*, Harvard Business Review, 2007.

8. Eytan Bakshy, *Everyone's an Influencer: Quantifying Influence on Twitter*, Proceedings of the Fourth ACM Conference, 2011.

9. Jesse Willms, *R3 Block chain Development Initiative Grows to 22 Banks Worldwide*, Bitcoin Magazine, 2015.

10. IBM Institute for Business Value, *Device Democracy: Saving the Future of the Internet of Things*, IBM Corporation, 2015.

11. Peter Thiel, *Zero to One: Notes on Startups, or How to Build the Future*, Crown Business, New York, 2014.

12. Julia La Roche, *Goldman Sachs Just Pulled a Silicon Valley Move*, Business Inside, 2015.

13. James Manyika, *Unlock the Potential of the Internet of Things*, McKinsey Global Institute, 2015.

14. Jeff Bezos, *Letter to shareholder*, 2014.

15. Larry Page, *Larry's Alphabet Letter*, 2016.

16. Julian Birkinshaw, *What to Expect from Agile*, MIT SMR 2018.

ACKNOWLEDGMENTS

Any book is impossible without the help of generous contributors. I thank all the business leaders and authors whose work is cited in the book, especially those who shared their experiences with me in detail about their struggling, surviving and thriving during digital transformations.

This book would not have happened without my gracious editor, Kelvin, providing invaluable feedback in crafting the tone and structure of the book, as well as sharpening each turn of phrase, tightening the prose, and ensuring that every idea would be clear to readers, with his professional, guidance, and kindness. He has my enduring gratitude.

Dean Zhang of ICCI Jiao Tong University, and Dean Herve of KEDGE business School, as my supervisors and instructors, supported my project at every stage from the very beginning. The fellow faculty members of my school provided both rich intellectual inspiration for many ideas and critical feedback within the writing process, helping me to focus on core messages and key opinions of this dynamic topic.

Lastly, I thank my husband, Jason Ma and my parents. They kept me going, encouraging me during the hard times I was totally engaged in writing. Their love and full support are the inspiration behind all my work.

FAN WU
New York City,   New York

ABOUT THE AUTHOR

Dr. Fan Wu is a general business manager, brand management professional, and marketing intelligence academic researcher, with more than 10 years of rich experience and inspiration in the beauty industry.

Her work experience includes the following: Deputy General Manager of HERBORIST Cosmetic Ltd in China and Oversea; Head of MKT of Elizabeth Arden Cosmetic Brand in China; Communication manager of EATON POWERWARE in the Asia-Pacific region; Head of Consulting for Digital CRM project of L'OREAL China LUXE Division, for the innovative digitalization and transformation of brands such as LANCOME, KIEHLS, BIOTHERM, YUSAI, SHU, YSL, GA, HR, etc.

Faye received her Ph.D. in communication management from Shanghai Jiao Tong University. She also holds an M.A. in Communication Art from New York Institute of Tech, and C.P.M. from New York University. Now she is teaching at USC-SJTU Institute of Culture and Creative Industry, and at KEDGE business school.

Her research fields include:

- Marketing & Branding Management
- Business Digital Transformation Management
- Digital Marketing & Data Intelligence